Universitext

Springer
New York
Berlin
Heidelberg
Barcelona
Hong Kong
London
Milan
Paris
Singapore
Tokyo

Universitext

Editors (North America): S. Axler, F.W. Gehring, and K.A. Ribet

Aksoy/Khamsi: Nonstandard Methods in Fixed Point Theory
Andersson: Topics in Complex Analysis
Aupetit: A Primer on Spectral Theory
Balakrishnan/Ranganathan: A Textbook of Graph Theory
Balser: Formal Power Series and Linear Systems of Meromorphic Ordinary
  Differential Equations
Bapat: Linear Algebra and Linear Models (2nd ed.)
Berberian: Fundamentals of Real Analysis
Booss/Bleecker: Topology and Analysis
Borkar: Probability Theory: An Advanced Course
Böttcher/Silbermann: Introduction to Large Truncated Toeplitz Matrices
Carleson/Gamelin: Complex Dynamics
Cecil: Lie Sphere Geometry: With Applications to Submanifolds
Chae: Lebesgue Integration (2nd ed.)
Charlap: Bieberbach Groups and Flat Manifolds
Chern: Complex Manifolds Without Potential Theory
Cohn: A Classical Invitation to Algebraic Numbers and Class Fields
Curtis: Abstract Linear Algebra
Curtis: Matrix Groups
DiBenedetto: Degenerate Parabolic Equations
Dimca: Singularities and Topology of Hypersurfaces
Edwards: A Formal Background to Mathematics I a/b
Edwards: A Formal Background to Mathematics II a/b
Foulds: Graph Theory Applications
Friedman: Algebraic Surfaces and Holomorphic Vector Bundles
Fuhrmann: A Polynomial Approach to Linear Algebra
Gardiner: A First Course in Group Theory
Gårding/Tambour: Algebra for Computer Science
Goldblatt: Orthogonality and Spacetime Geometry
Gustafson/Rao: Numerical Range: The Field of Values of Linear Operators
  and Matrices
Hahn: Quadratic Algebras, Clifford Algebras, and Arithmetic Witt Groups
Holmgren: A First Course in Discrete Dynamical Systems
Howe/Tan: Non-Abelian Harmonic Analysis: Applications of $SL(2, R)$
Howes: Modern Analysis and Topology
Hsieh/Sibuya: Basic Theory of Ordinary Differential Equations
Humi/Miller: Second Course in Ordinary Differential Equations
Hurwitz/Kritikos: Lectures on Number Theory
Jennings: Modern Geometry with Applications
Jones/Morris/Pearson: Abstract Algebra and Famous Impossibilities
Kannan/Krueger: Advanced Analysis
Kelly/Matthews: The Non-Euclidean Hyperbolic Plane
Kostrikin: Introduction to Algebra
Luecking/Rubel: Complex Analysis: A Functional Analysis Approach
MacLane/Moerdijk: Sheaves in Geometry and Logic
Marcus: Number Fields
McCarthy: Introduction to Arithmetical Functions

(continued after index)

R. Balakrishnan   K. Ranganathan

A Textbook of
Graph Theory

With 200 Figures

 Springer

R. Balakrishnan
Department of Mathematics
Bharathidasan University
Tiruchirappalli
Tamil Nadu 620 024
India

K. Ranganathan
National College
Tiruchirappalli
Tamil Nadu 620 001
India

Mathematics Subject Classification (1991): 05-01, 05Cxx

Library of Congress Cataloging-in-Publication Data
Balakrishnan, R.
    A textbook of graph theory / R. Balakrishnan, K. Ranganathan.
      p.   cm.—(Universitext)
    Includes bibliographical references and index.
    ISBN 0-387-98859-9 (alk. paper)
    1. Graph theory.   I. Ranganathan, K.   II. Title.
  QA166.B25   1999
  511'.5—dc21   99-15016

Printed on acid-free paper.

Production managed by Timothy Taylor; manufacturing supervised by Jacqui Ashri.
Typeset by TechBooks, Fairfax, VA.
Printed and bound by R.R. Donnelley and Sons, Harrisonburg, VA.
Printed in the United States of America.

9 8 7 6 5 4 3 2 1

ISBN 0-387-98859-9 Springer-Verlag New York Berlin Heidelberg   SPIN 10727816

Dedicated to our parents

Preface

Graph theory has witnessed an unprecedented growth in the twentieth century. The best barometer to indicate this growth is the explosion in the number of pages that Section 05: Combinatorics (in which the major share is taken by graph theory) occupies in the *Mathematical Reviews*. One of the main reasons for this growth is the applicability of graph theory in many other disciplines such as physics, chemistry, psychology, and sociology. Yet another reason is that some of the problems in theoretical computer science that deal with complexity can be transformed into graph-theoretical problems.

This book aims to provide a good background in the basic topics of graph theory. It does not presuppose deep knowledge of any branch of mathematics. As a basic text in graph theory, it contains, for the first time, Dirac's theorem on k-connected graphs (with adequate hints), Harary-Nashwilliams's theorem on the hamiltonicity of line graphs, Toida-McKee's characterization of Eulerian graphs, the Tutte matrix of a graph, David Sumner's result on claw-free graphs, Fournier's proof of Kuratowski's theorem on planar graphs, the proof of the nonhamiltonicity of the Tutte graph on 46 vertices, and a concrete application of triangulated graphs.

An ambitious teacher can cover the entire book in a one-year (equivalent to two semesters) master's course in mathematics or computer science. However, a teacher who wants to proceed at a leisurely pace can omit the sections that are starred. Exercises that are starred are nonroutine.

The book can also be adapted for an undergraduate course in graph theory by selecting the following sections: 1.0–1.5, 2.0–2.2, 3.0–3.3, 4.0–4.4, 5.0–5.3, 5.4 (omitting consequences of Hall's theorem), 5.5 (omitting the Tutte matrix), 6.0–6.2, 7.0–7.1, 7.4 (omitting Vizing's theorem), 7.7, 8.0–8.3, and chapter 10.

Several people have helped us by reviewing the manuscript in parts and offering constructive suggestions: S. Arumugam, S. A. Choudum, P. K. Jha, P. Paulraja, G. Ramachandran, S. Ramachandran, G. Ravindra, E. Sampathkumar, and R. Sampathkumar. We thank all of them most profusely for their kind gesture in sparing for our sake a portion of their precious time. Our special thanks are due to P. Paulraja and R. Samptathkumar, who had been a constant source of inspiration to us ever since we started working on this book rather seriously. We also thank D. Kannan, Department of Mathematics, University of Georgia, for reading the manuscript and suggesting some stylistic changes.

We also take this opportunity to thank the authorities of our institutions, Annamalai University, Annamalainagar, and National College, Tiruchirapalli, for their kind encouragement. Finally, we thank the University Grants Commission, Government of India, for their financial support for writing this book.

Our numbering scheme for theorems and exercises is as follows. Each exercise bears two numbers, whereas each theorem, lemma, and so forth bears three numbers. Therefore, Exercise 2.4 is the fourth exercise of section 2 of a particular chapter, and Theorem 6.5.1 is the first result of section 5 of chapter 6.

Tamil Nadu, India   R. BALAKRISHNAN
                                                     K. RANGANATHAN

Contents

I
Basic Results

1.0. Introduction

Graphs serve as mathematical models to analize successfully many concrete real-world problems. Certain problems in physics, chemistry, communications science, computer technology, genetics, psychology, sociology, and linguistics can be formulated as problems in graph theory. Also many branches of mathematics, such as group theory, matrix theory, probability, and topology, have interactions with graph theory.

Some puzzles and various problems of a practical nature have been instrumental in the development of various topics in graph theory. The famous Königsberg Bridge problem has been the inspiration for the development of Eulerian graph theory. The challenging Hamiltonian graph theory has been developed from the "Around the World" game of Sir William Hamilton. The theory of acyclic graphs was developed for solving problems of electrical networks, and the study of "trees" was developed for enumerating isomers of organic compounds. The well-known four-color problem has formed the very basis for the development of planarity in graph theory and combinatorial topology. Problems of linear programming and operations research (such as maritime traffic problems) can be tackled by the theory of flows in networks. Kirkman's schoolgirls problem and scheduling problems are examples of problems that can be solved by graph colorings. The study of simplicial complexes can be associated with the study of graph theory. Many more such problems can be added to this list.

1.1. Basic Concepts

Consider a road network of a town consisting of streets and street intersections. Figure 1.1(a) represents the road network of a city. Figure 1.1(b) denotes the corresponding graph of this network, where the street intersections are represented by points, and the streets joining any pair of intersections are represented by arcs (not necessarily straight lines). The road network of Figure 1.1 is a typical example of a graph in which intersections and streets are, respectively, the "vertices" and "edges" of the graph. (Note that in the road network of Figure 1.1(a) there are two streets joining the intersections J_7 and J_8, and there is a loop street starting and ending at J_2.)

We present here a formal definiton of a graph.

Definition 1.1.1 A *graph* is an ordered triple $G = (V(G), E(G), I_G)$, where $V(G)$ is a nonempty set, $E(G)$ is a set disjoint from $(V(G)$, and I_G is an "incidence" map that associates with each element of $E(G)$, an unordered pair of elements (same or distinct) of $V(G)$. Elements of $V(G)$ are called the *vertices* (or *nodes* or

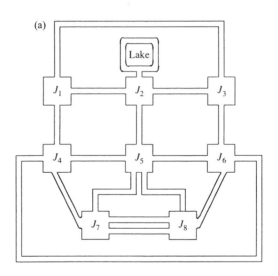

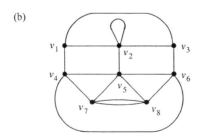

FIGURE 1.1. (a) A road network and (b) the graph corresponding to the road network in (a)

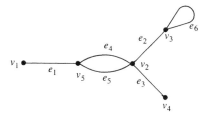

FIGURE 1.2. Graph $(V(G), E(G), I_G)$ described in Example 1.1.2

points) of G, and elements of $E(G)$ are called the *edges* (or *lines*) of G. If, for the edge e of G, $I_G(e) = \{u, v\}$, we write $I_G(e) = uv$.

Example 1.1.2   If $V(G) = \{v_1, v_2, v_3, v_4, v_5\}$, $E(G) = \{e_1, e_2, e_3, e_4, e_5, e_6\}$ and I_G is given by $I_G(e_1) = \{v_1, v_5\}$, $I_G(e_2) = \{v_2, v_3\}$, $I_G(e_3) = \{v_2, v_4\}$, $I_G(e_4) = \{v_2, v_5\}$, $I_G(e_5) = \{v_2, v_5\}$, $I_G(e_6) = \{v_3, v_3\}$, then $(V(G), E(G), I_G)$ is a graph (see Figure 1.2).

Diagramatic Representation of a Graph 1.1.3   Each graph can be represented by a diagram in the plane. In this diagram, each vertex of the graph is represented by a point, with distinct vertices being represented by distinct points. Each edge is represented by a simple "Jordan" arc joining two (not necessarily distinct) vertices. The diagrammatic representation of a graph aids in visualizing many concepts relating to graphs and the systems of which they are models. In a diagrammatic representation of a graph, it is possible that two edges interesect at a point that is not necessarily a vertex of the graph.

Definition 1.1.4   If $I_G(e) = \{u, v\}$, then the vertices u and v are called the *end vertices* or *ends* of the edge e. Each edge is said to join its ends; in this case, we say that e is *incident* with each one of its ends. Also, the vertices u and v are then incident with e. A set of two or more edges of a graph G is called a set of *multiple* or *parallel edges* if they have the same ends. If e is the only edge with end vertices u and v, we write $e = uv$. An edge for which the two ends are the same is called a *loop* at the common vertex. A vertex u is a *neighbor* of v in G, if uv is an edge of G, and $u \neq v$. The set of all neighbors of v is the *open neighborhood* of v or the neighbor set of v, and is denoted by $N(v)$; the set $N[v] = N(v) \cup \{v\}$ is the *closed neighborhood* of v in G. When G must be explicit, these open and closed neighborhoods are denoted by $N_G(v)$ and $N_G[v]$, respectively. Vertices u and v are *adjacent* to each other in G if, and only if, there is an edge of G with u and v as its ends. Two distinct edges e and f are said to be *adjacent* if, and only if, they have a common end vertex. A graph is *simple* if it has no loops and no multiple edges. Thus for a simple graph G, the incidence function I_G is one-to-one. Hence, an edge of a simple graph is identified with the pair of its ends. A simple graph therefore may be considered as an ordered pair $(V(G), E(G))$, where $V(G)$ is a nonempty set and $E(G)$ is a set of unordered pairs of elements of $V(G)$ (each edge of the graph being identified with the pair of its ends).

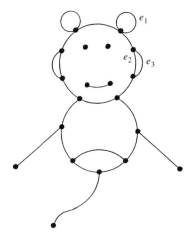

FIGURE 1.3. A graph diagram; e_1 is a loop and $\{e_2, e_3\}$ is a set of multiple edges

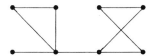

FIGURE 1.4. A simple graph

Example 1.1.5 In the graph of Figure 1.2, edge, $e_3 = v_2 v_4$, edges e_4 and e_5 form multiple edges, e_6 is a loop at v_3, $N(v_2) = \{v_3, v_4, v_5\}$, $N(v_3) = \{v_2\}$, $N[v_2] = \{v_2, v_3, v_4, v_5\}$ and $N[v_2] = N(v_2) \cup \{v_2\}$. Further, v_2 and v_5 are adjacent vertices, and e_3 and e_4 are adjacent edges.

Definition 1.1.6 A graph is called *finite* if both $V(G)$ and $E(G)$ are finite. A graph that is not finite is called *infinite*. Unless otherwise stated, all graphs in this text are finite. Throughout this book, we denote by $n(G)$ and $m(G)$ the number of vertices and edges of the graph G, respectively. The number $n(G)$ is called the *order of G* and $m(G)$ is the *size of G*. When explicit reference to the graph G is not needed, $V(G)$, $E(G)$, $n(G)$, and $m(G)$ will be denoted simply by V, E, n, and m, respectively.

Figure 1.3 is a graph with loops and multiple edges; Figure 1.4 represents a simple graph.

Remark 1.1.7 The representation of graphs on other surfaces, such as a sphere, a torus, or a Möbius band, could also be considered. Often a diagram of a graph is identified with the graph itself.

Definition 1.1.8 A graph is said to be *labeled*, if its n vertices are distinguished from one another by labels such as $v_1, v_2, \ldots, v_n$ (Figure 1.5).

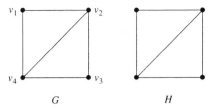

FIGURE 1.5. A labeled graph (G) and an unlabeled graph (H)

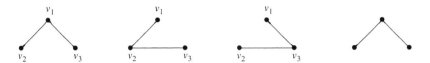

FIGURE 1.6. Labeled and unlabeled simple graphs on three vertices

Note that there are three different labeled simple graphs, on three vertices each having two edges, whereas there is only one unlabeled simple graph of the same order and size (see Figure 1.6).

Isomorphism of Graphs 1.1.9   A graph isomorphism, which we now define, is a concept similar to isomorphism in algebraic structures. Let $G = (V(G), E(G), I_G)$ and $H = (V(H), E(H), I_H)$ be two graphs. A *graph isomorphism* from G to H (written $G \cong H$) is a pair (ϕ, θ), where $\phi : V(G) \rightarrow V(H)$ and $\theta : E(G) \rightarrow E(H)$ are bijections with the property that $I_G(e) = \{u, v\}$ if, and only if, $I_H(\theta(e)) = \{\phi(u), \phi(v)\}$. If (ϕ, θ) is a graph isomorphism, the pair of inverse mappings (ϕ^{-1}, e^{-1}) is also a graph isomorphism. Note that the bijection ϕ satisfies the condition that *u and v are end vertices of an edge e of G if, and only if, $\phi(u)$ and $\phi(v)$ are end vertices of the edge $\theta(e)$ in H*.

Simple Graphs and Isomorphisms 1.1.10   If graphs G and H are simple, a bijection $\phi : V(G) \rightarrow V(H)$ such that u and v are adjacent in G if, and only if, $\phi(u)$ and $\phi(v)$ are adjacent in H induces a bijection $\theta : E(G) \rightarrow E(H)$ satisfying the condition that $I_G(e) = \{u, v\}$ if, and only if, $I_H(\theta(e)) = \{\phi(u), \phi(v)\}$. Hence ϕ itself is referred to as an isomorphism in the case of simple graphs G and H. Thus if G and H are simple graphs, an isomorphism from G to H is a bijection $\phi : V(G) \rightarrow V(H)$ such that u and v are adjacent in G if, and only if, $\phi(u)$ and $\phi(v)$ are adjacent in H. Figure 1.7 exhibits two isomorphic graphs P and H, where P is the Petersen graph.

Exercise 1.1   Let G and H be simple graphs and let $\phi : V(G) \rightarrow V(H)$ be a bijection such that $uv \in E(G)$ implies that $\phi(u)\phi(v) \in E(H)$. Show, by means of an example, that ϕ need not be an isomorphism from G to H.

Definition 1.1.11   A simple graph G is said to be *complete* if every pair of distinct vertices of G are adjacent in G. Any two complete graphs each on a set of n vertices are isomorphic; each such graph is denoted by K_n.

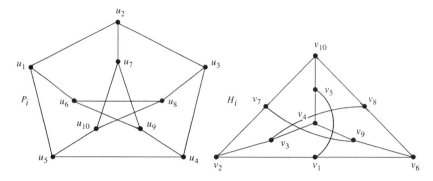

FIGURE 1.7. Isomorphic graphs

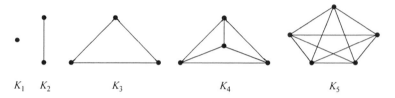

K_1 K_2 K_3 K_4 K_5

FIGURE 1.8. Some complete graphs

FIGURE 1.9. A totally disconnected graph on five vertices

A simple graph with n vertices can have at most $\binom{n}{2} = n(n-1)/2$ edges. K_n has the maximum number of edges among all simple graphs with n vertices. On the other extreme, a graph may possess no edge at all. Such a graph is called a *totally disconnected graph* (Figure 1.9). Thus, *for a simple graph G with n vertices, we have* $0 \le m(G) \le n(n-1)/2$.

Definitions 1.1.12   A graph is *trivial* if its vertex set is a singleton and it contains no edges. A graph is *bipartite* if its vertex set can be partitioned into two nonempty subsets X and Y such that each edge of G has one end in X and the other in Y. The pair (X, Y) is called a *bipartition* of the bipartite graph. The bipartite graph G with bipartition (X, Y) is denoted by $G(X, Y)$. A simple bipartite graph $G(X, Y)$ is *complete* if each vertex of X is adjacent to all the vertices of Y. If $G(X, Y)$ is complete with $|X| = p$ and $|Y| = q$, then $G(X, Y)$ is denoted by $K_{p,q}$. A complete bipartite graph of the form $K_{1,q}$ is called a *star*. (See Figure 1.10.)

Definition 1.1.13   Let G be a simple graph. Then the *complement* G^c of G is defined by taking $V(G^c) = V(G)$ and making two vertices u and v adjacent in G^c

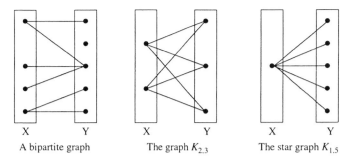

X Y	X Y	X Y
A bipartite graph	The graph $K_{2,3}$	The star graph $K_{1,5}$

FIGURE 1.10. Bipartite graphs

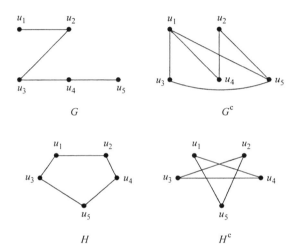

FIGURE 1.11. Two simple graphs and their complements

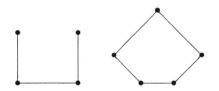

FIGURE 1.12. Self-complementary graphs

if, and only if, they are nonadjacent in G (see Figure 1.11). It is clear that G^c *is also a simple graph and that* $(G^c)^c = G$.

If $|V(G)| = n$, then clearly, $|E(G)| + |E(G^c)| = |E(K_n)| = n(n-1)/2$.

Definition 1.1.14 A simple graph G is called *self-complementary* if $G \cong G^c$.

For example, the graphs shown in Figure 1.12 are self-complementary.

Exercise 1.2 Find the complement of the following simple graph:

1.2. Subgraphs

Definitions 1.2.1 A graph H is called a *subgraph* of G if $V(H) \subseteq V(G)$, $E(H)$ $\subseteq E(G)$, and I_H is the restriction of I_G to $E(H)$. If H is a subgraph of G, then G is said to be a *supergraph* of H. A subgraph H of a graph G is a *proper subgraph* of G if either $V(H) \neq V(G)$ or $E(H) \neq E(G)$. (Hence when G is given, for any subgraph H of G, the incidence function is already determined so that H can be specified by its vertex and edge sets.) A subraph H of G is said to be an *induced subgraph* of G if each edge of G having its ends in $V(H)$ is also an edge of H. A subgraph H of G is a *spanning subgraph* of G, if $V(H) = V(G)$. The induced subgraph of G with vertex set $S \subseteq V(G)$ is called the *subgraph of G induced by S* and is denoted by $G[S]$. Let E' be a subset of E and let S denote the subset of V consisting of all the end vertices in G of edges in E'. Then the graph $(S, E', I_G|E')$ is the *subgraph of G induced by the edge set E' of G*. It is denoted by $G[E']$. (See Figure 1.13.) Let u and v be vertices of a graph G. By $G + uv$, we mean the graph obtained by adding a new edge uv to G.

Definitions 1.2.2 A *clique* of G is a complete subgraph of G. A clique of G is a *maximal clique* of G if it is not properly contained in another clique of G. (See Figure 1.13.)

Definition 1.2.3 *Deletion of vertices and edges in a graph*: Let G be a graph, S a proper subset of the vertex set V, and E' a subset of E. The subgraph $G[V - S]$ is said to be obtained from G by the *deletion* of S. This subgraph is denoted by $G - S$. If $S = \{v\}$, $G - S$ is simply denoted by $G - v$. The spanning subgraph of

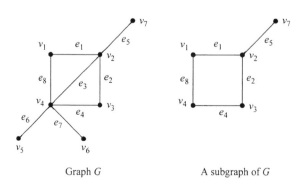

Graph G A subgraph of G

FIGURE 1.13. Various subgraphs and cliques of G

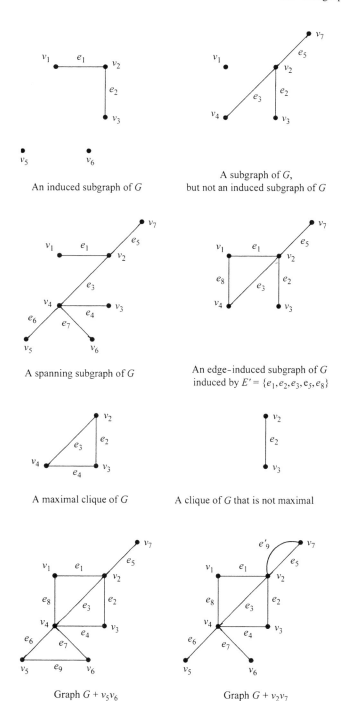

An induced subgraph of G

A subgraph of G,
but not an induced subgraph of G

A spanning subgraph of G

An edge-induced subgraph of G
induced by $E' = \{e_1, e_2, e_3, e_5, e_8\}$

A maximal clique of G

A clique of G that is not maximal

Graph $G + v_5 v_6$

Graph $G + v_2 v_7$

FIGURE 1.13 (*Continued*)

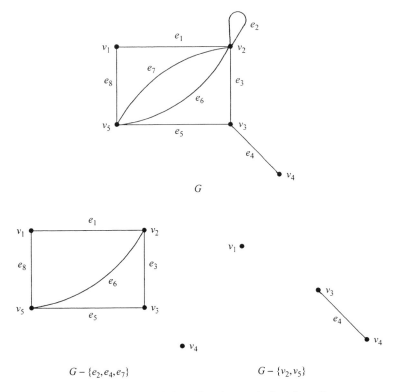

FIGURE 1.14. Deletion of vertices and edges from G

G with the edge set E/E' is the subgraph obtained from G by deleting the edge subset E'. This subgraph is denoted by $G - E'$. Whenever $E' = \{e\}$, $G - E'$ is simply denoted by $G - e$. Note that when a vertex is deleted from G, all the edges incident to it are also deleted from G, whereas the deletion of an edge from G does not affect the vertices of G. (See Figure 1.14.)

1.3. Degrees of Vertices

Definition 1.3.1 Let G be a graph and $v \in V$. The number of edges incident at v in G is called the *degree* (or *valency*) of the vertex v in G and is denoted by $d_G(v)$, or simply $d(v)$ when G requires no explicit reference. A loop at v is to be counted twice in computing the degree of v. The minimum (respectively, maximum) of the degrees of the vertices of a graph G is denoted $\delta(G)$ or δ (respectively, $\Delta(G)$ or Δ). A graph G is called *k-regular*, if every vertex of G has degree k. A graph is said to be *regular* if it is k-regular for some nonnegative integer k. In particular, a 3-regular graph is called a *cubic graph*.

Definition 1.3.2 A spanning 1-regular subgraph of G is called a *1-factor* or a *perfect matching* of G. For example, in the graph G of Figure 1.15 each of the pairs $\{ab, cd\}$ and $\{ad, bc\}$ is a 1-factor of G.

FIGURE 1.15. Graph with 1-factors

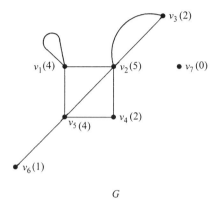

G

FIGURE 1.16. Degress of vertices of graph G

Definition 1.3.3   A vertex of degree 0 is known as an *isolated vertex* of G. A vertex of degree 1 is called a *pendant vertex* of G, whereas the unique edge of G incident to such a vertex of G is a *pendant edge* of G. A sequence formed by the degrees of vertices of G is called a *degree sequence* of G. It is customary to give this sequence in the nonincreasing or nondecreasing order.

In the graph G of Figure 1.16, the numbers within the parentheses indicate the degrees of the corresponding vertices. In G, v_7 is an isolated vertex, v_6 is a pendant vertex, and $v_5 v_6$ is a pendant edge. The degree sequence of G is $(0, 1, 2, 2, 4, 4, 5)$.

The very first theorem of graph theory was due to Leonhard Euler (1707–1783). This theorem connects the degrees of vertices and the number of edges of a graph.

Theorem 1.3.4 (Euler)   *The sum of the degrees of the vertices of a graph is equal to twice the number of its edges.*

Proof   If $e = uv$ is an edge of G, e is counted once while counting the degrees of each of u and v (even when $u = v$). Hence each edge contributes 2 to the sum of the degrees of the vertices. Thus the m edges of G contribute $2m$ to the degree sum. □

Remark 1.3.5   If $d = (d_1, d_2, \ldots, d_n)$ is the degree sequence of G, then the above theorem gives the equation $\sum_{i=1}^{n} d_i = 2m$, where n and m are the order and size of G, respectively.

Corollary 1.3.6   *In any graph G, the number of vertices of odd degrees is even.*

Proof Let V_1 and V_2 be the subsets of vertices of G with odd and even degrees, respectively. By Theorem 1.3.4,

$$2m(G) = \sum_{v \in V} d_G(v) = \sum_{v \in V_1} d_G(v) + \sum_{v \in V_2} d_G(v).$$

As $2m(G)$ and $\sum_{v \in V_2} d_G(v)$ are even, $\sum_{v \in V_1} d_G(v)$ is even. Since for each $v \in V_1$, $d_G(v)$ is odd, $|V_1|$ must be even.   □

Exercise 3.1 Show that if G and H are isomorphic graphs, then each pair of corresponding vertices of G and H have the same degree.

Exercise 3.2 Let $(d_1, d_2, \ldots, d_n)$ be the degree sequence of a graph, and r be any positive integer. Show that $\sum_{i=1}^{n} d_i^r$ is even.

Definition 1.3.7 *Graphical sequences:* A sequence of nonnegative integers $d = (d_1, d_2, \ldots, d_n)$ is called *graphical* if there exists a simple graph whose degree sequence is d. Clearly, a necessary condition for $d = (d_1, d_2, \ldots, d_n)$ to be graphical is that $\sum_{i=1}^{n} d_i$ is even and $d_i \geq 0, 1 \leq i \leq n$. These conditions, however, are not sufficient, as Example 1.3.8 shows.

Example 1.3.8 The sequence $d = (7, 6, 3, 3, 2, 1, 1, 1)$ is not graphical, even though each term of d is a nonnegative integer and the sum of the terms is even. Indeed, if d were graphical, there must exist a simple graph G with eight-vertices whose degree sequence is d. Let v_0 and v_1 be the vertices of G whose degrees are 7 and 6, respectively. Since G is simple, v_0 is adjacent to all the remaining vertices of G, and v_1, besides v_0, should be adjacent to another five vertices. This means that in $V - \{v_0, v_1\}$ there must be at least five vertices of degree at least 2. But this is not the case.   □

Exercise 3.3 If $d = (d_1, d_2, \ldots, d_n)$ is any sequence of nonnegative integers with $\sum_{i=1}^{n} d_i$ even, show that there exists a graph (not necessarily simple) with d as its degree sequence.

Example 1.3.9 In any group of n persons ($n \geq 2$), there are at least two with the same number of friends.

Solution Denote the n persons by $v_1, v_2, \ldots, v_n$. Let G be the simple graph with vertex set $V = \{v_1, v_2, \ldots, v_n\}$ in which v_i and v_j are adjacent if, and only if, the corresponding persons are friends. Then the number of friends of v_i is just the degree of v_i in G. Hence , to solve the problem, we must prove that there are two vertices in G with the same degree. If this were not the case, the degrees of the vertices of G must be $0, 1, 2, \ldots, (n-1)$ in some order. However, a vertex of degree $(n-1)$ must be adjacent to all the other vertices of G, and consequently there cannot be a vertex of degree 0 in G. This contradiction shows that the degrees of the vertices of G cannot all be distinct, and hence at least two of them should have the same degree. □

Exercise 3.4 Let G be a graph with n vertices and m edges. Assume that each vertex of G is of degree either k or $k + 1$. Show that the number of vertices of degree k in G is $(k + 1)n - 2m$.

1.4. Paths and Connectedness

Definitions 1.4.1 A *walk* in a graph G is an alternating sequence $W : v_0 e_1 v_1 e_2 v_2 \ldots e_n v_n$ of vertices and edges beginning and ending with vertices in which v_{i-1} and v_i are the ends of e_i; v_0 is the *origin* and v_n is the *terminus* of W. The walk W is said to join v_0 and v_n; it is also referred to as a $v_0 - v_n$ walk. If the graph is simple, a walk is determined by the sequence of its vertices. The walk is *closed* if $v_0 = v_n$ and is *open* otherwise. A walk is called a *trail* if all the edges appearing in the walk are distinct. It is called a *path* if all the vertices are distinct. Thus a path in G is automatically a trail in G. When writing a path, we usually omit the edges. A *cycle* is a closed trail in which the vertices are all distinct. The *length* of a walk is the number of edges in it. A walk of length zero consists of just a single vertex.

Example 1.4.2 In the graph of Figure 1.17, $v_5 e_7 v_1 e_1 v_2 e_4 v_4 e_5 v_1 e_7 v_5 e_9 v_6$ is a walk but not a trail; $v_1 e_1 v_2 e_2 v_3 e_3 v_2 e_1 v_1$ is a closed walk; $v_1 e_1 v_2 e_4 v_4 e_5 v_1 e_7 v_5$ is a trail; $v_6 e_8 v_1 e_1 v_2 e_2 v_3$ is a path and $v_1 e_1 v_2 e_4 v_4 e_6 v_5 e_7 v_1$ is a cycle. Also $v_6 v_1 v_2 v_3$ is a path and $v_1 v_2 v_4 v_5 v_6 v_1$ is a cycle of the above graph.

Definition 1.4.3 A graph that is a cycle of length n is denoted by C_n. P_n denotes a path on n vertices. In particular, C_3 is often referred to as a *triangle*, C_4 as a *square*, and C_5 as a *pentagon*. If $P = v_0 e_1 v_1 e_2 v_2 \ldots e_n v_n$ is a path, then $P^{-1} = v_n e_n v_{n-1} e_{n-1} v_{n-2} \ldots v_1 e_1 v_0$ is also a path and P^{-1} is called the *inverse* of the path P. The subsequence $v_i e_{i+1} v_{i+1} \ldots e_j v_j$ of P is called the $v_i - v_j$ *section* of P.

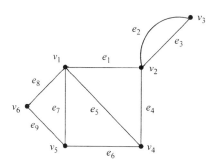

FIGURE 1.17. Graph illustrating walks, trails, paths, and cycles

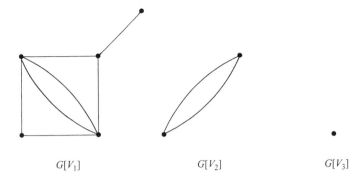

$G[V_1]$ $\qquad\qquad\qquad$ $G[V_2]$ $\qquad\qquad\qquad$ $G[V_3]$

FIGURE 1.18. A graph G with three components

Definition 1.4.4   Let G be a graph. Two vertices u and v of G are said to be *connected* if there is a u–v path in G. The relation "connected" is an equivalence relation on $V(G)$. Let $V_1, V_2, \ldots, V_\omega$ be the equivalence classes. The subgraphs $G[V_1], G[V_2], \ldots, G[V_\omega]$ are called the *components* of G. If $\omega = 1$, the graph is called *connected*; otherwise, the graph is *disconnected* with ω components. (see Figure 1.18.)

Remark 1.4.5   The components of G are clearly the maximal connected sub-graphs of G. We denote the number of components of G by $\omega(G)$. Let u and v be two vertices of G. If u and v are in the same component of G, we define $d(u, v)$ to be the length of a shortest u–v path in G; otherwise we define $d(u, v)$ to be ∞. If G is a connected graph, then d is a distance function or metric on $V(G)$; that is, $d(u, v)$ satisfies the following conditions:
  (i) $d(u, v) \geq 0$, and $d(u, v) = 0$ if, and only if, $u = v$,
  (ii) $d(u, v) = d(v, u)$ and
  (iii) $d(u, v) \leq d(u, w) + d(w, v)$, for every w in $V(G)$.

Exercise 4.1   Prove that the function d defined above is indeed a metric on $V(G)$.

Exercise 4.2   In the following graph, find a closed trail of length 7 that is not a cycle:

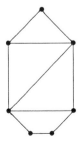

We now give some results relating to connectedness of graphs.

Proposition 1.4.6   *If G is simple and $\delta \geq \frac{n-1}{2}$, then G is connected.*

Proof   Assume the contrary. Then G has at least two components, say G_1, G_2. Let v be any vertex of G_1. As $\delta \geq \frac{n-1}{2}, d(v) \geq \frac{n-1}{2}$. All the vertices adjacent to v in G must belong to G_1. Hence G_1 contains at least $d(v) + 1 \geq \frac{n-1}{2} + 1 = \frac{n+1}{2}$ vertices. Similarly G_2 contains at least $\frac{n+1}{2}$ vertices. Therefore, G has at least $\frac{n+1}{2} + \frac{n+1}{2} = n + 1$ vertices, which is a contradiction.   □

Exercise 4.3   Give an example of a nonsimple disconnected graph with $\delta \geq \frac{n-1}{2}$.

Exercise 4.4   Show by means of an example that the condition $\delta \geq \frac{n-2}{2}$ for a simple graph G need not imply that G is connected.

Exercise 4.5   In a group of six people, prove that there must be three people who are mutually acquainted or three people who are mutually nonacquainted.

Our next result shows that of the two graphs G and G^c, at least one of them is connected.

Theorem 1.4.7   *If a simple graph G is not connected, then G^c is connected.*

Proof   Let u and v be any two vertices of G^c (and therefore of G). If u and v belong to different components of G, then obviously u and v are nonadjacent in G and so they are adjacent in G^c. Thus u and v are connected in G^c. In case u and v belong to the same component of G, take a vertex w of G not belonging to this component of G. Then uw and vw are not edges of G and hence they are edges of G^c. Then uwv is a u–v path in G^c. Thus G^c is connected.   □

Exercise 4.6   Show that if G is a self-complementary graph of order n, then $n \equiv 0$ or $1 \pmod 4$.

Exercise 4.7   Show that if a self-complementary graph contains a pendant vertex, then it must have at least another pendant vertex.

The next theorem gives an upper bound on the number of edges in a simple graph.

Theorem 1.4.8   *The number of edges of a simple graph with ω components cannot exceed $\frac{(n-\omega)(n-\omega+1)}{2}$.*

Proof   Let $G_1, G_2, \ldots, G_\omega$ be the components of a simple graph G and let n_i be the number of vertices of $G_1, 1 \leq i \leq \omega$. Then $m(G_i) \leq \frac{n_i(n_i-1)}{2}$ and hence $m(G) \leq \sum_{i=1}^{\omega} \frac{n_i(n_i-1)}{2}$. Since $n_i \geq 1$ for each i, $1 \leq i \leq \omega$, $n_i = n - (n_1 + \cdots +$

$n_{i-1} + n_{i+1} + \cdots + n_\omega) \leq n - \omega + 1$. Hence $\sum_{i=1}^{\omega} \frac{n_i(n_i - 1)}{2} \leq \sum_{i=1}^{\omega} \frac{(n-\omega+1)(n_i-1)}{2} = \frac{(n-\omega+1)}{2} \sum_{i=1}^{\omega} (n_i - 1) = \frac{(n-\omega+1)}{2} \left[\left(\sum_{i=1}^{\omega} n_i \right) - \omega \right] = \frac{(n-\omega+1)(n-\omega)}{2}$.   □

Definition 1.4.9   A graph G is called *locally connected* if for every vertex v of G, the neighbor set of v in G, $N_G(v)$, is connected.

A cycle is *odd* or *even* according as its length is odd or even. We now characterize bipartite graphs.

Theorem 1.4.10   *A graph is bipartite if, and only if, it contains no odd cycles.*

Proof   Suppose that G is a bipartite graph with the bipartition (X, Y). Let $C = v_1 e_1 v_2 e_2 v_3 e_3 \ldots v_k e_k v_1$ be a cycle in G. Then $k \geq 2$. Without loss of generality we can suppose that $v_1 \in X$. As v_2 is adjacent to v_1, $v_2 \in Y$. Similarly v_3 belongs to X, v_4 to Y, and so on. Thus, $v_i \in X$ or Y according as i is odd or even, $1 \leq i \leq k$. Since $v_k v_1$ is an edge of G and $v_1 \in X$, $v_k \in Y$. Accordingly, k is even and C is an even cycle.

Conversely, let us suppose that G contains no odd cycles. We first assume that G is connected. Let u be a vertex of G. Define $X = \{v \in V \mid d(u, v)$ is even$\}$ and $Y = \{v \in V \mid d(u, v)$ is odd$\}$. We will prove that (X, Y) is a bipartition of G. To prove this we have only to show that no two vertices of X as well as of Y are adjacent in G. Let v, w be two vertices of X. Then $p = d(u, v)$ and $q = d(u, w)$ are even. Further, there exists a u–v path P of length p and a u–w path Q of length q in G. (See Figure 1.19.) Evidently, P and Q are the shortest paths in G. Let w_1 be a vertex common to P and Q such that the w_1–v section of P and the w_1–w section of Q contain no vertices common to P and Q. Then the u–w_1 sections of P and Q have the same length.

Hence the lengths of the w_1–v section of P and the w_1–w section of Q are both even or both odd. Now, if $e = vw$ is an edge of G, then the w_1–v section of P followed by the edge vw and the w–w_1 section of the w–u path Q^{-1} is an odd cycle in G, contradicting the hypothesis. This contradiction proves the result when G is connected.

If G is not connected, let $G_1, G_2, \ldots, G_\omega$ be the components of G. By hypothesis, no component of G contains an odd cycle. Hence by the previous paragraph, each component G_i, $1 \leq i \leq \omega$, is bipartite. Let (X_i, Y_i) be the bipartition of G_i.

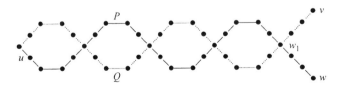

FIGURE 1.19. Graph for proof of Theorem 1.4.10

Then (X, Y), where $X = \bigcup_{i=1}^{\omega} X_i$ and $Y = \bigcup_{i=1}^{\omega} Y_i$, is a bipartition of G, and G is a bipartite graph. □

Exercise 4.8 Prove that a simple nontrivial graph G is connected if, and only if, for any partition of V into two nonempty subsets V_1 and V_2, there is an edge joining a vertex of V_1 to a vertex of V_2.

Example 1.4.11 Prove that, in a connected graph G with at least three vertices, any two longest paths have a vertex in common.

Solution Suppose $P = u_1 u_2 \ldots u_k$ and $Q = v_1 v_2 \ldots v_k$ are two longest paths in G having no vertex in common. As G is connected, there exists a u_1-v_1 path P' in G. Certainly there exist vertices u_r and v_s of P', $1 \leq r \leq k, 1 \leq s \leq k$ such that the u_r-v_s section P'' of P' has no internal vertex in common with P or Q.
 Now, of the two sections u_1-u_r and u_r-u_k of P, one must have length at least $k/2$. Similarly, of the two sections v_1-v_s and v_s-v_k of Q, one must have length at least $k/2$. Let these sections be P_1 and Q_1, respectively. Then $P_1 \cup P'' \cup Q_1$ is a path of length at least $(k/2) + (k/2) + 1$, contradicting that k is the length of a longest path in G. (See Figure 1.20.) □

Exercise 4.9 Prove that, in a simple graph G, the union of two distinct paths joining two distinct vertices contains a cycle.

Exercise 4.10 Show by means of an example that the union of two distinct walks joining two distinct vertices of a simple graph G need not contain a cycle.

Exercise 4.11 If a simple connected graph G is not complete, then prove that there exist three vertices u, v, w of G such that uv and vw are edges of G, but uw is not an edge of G.

Exercise 4.12 (see reference: [111]) Show that a simple connected graph G is complete if, and only if, for some vertex v of G, $N[v] = N[u]$ for every $u \in N[v]$.

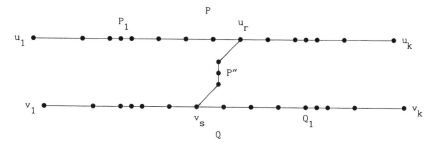

FIGURE 1.20. Graph for Solution 1.4.11

Exercise 4.13 A simple graph G is called *highly irregular* if for each $v \in V(G)$, the degrees of the neighbors of v are all distinct. For example, P_4 is a graph with this property. Prove that there exist no connected highly irregular graphs of orders 3 and 5.

Example 1.4.12 If G is simple and $\delta \geq k$, then G contains a path of length at least k.

Solution Let $P = v_0 v_1 \ldots v_r$ be a longest path in G. Then the vertices adjacent to v_r can only be from among $v_0, v_1, \ldots, v_{r-1}$. Hence the length of $P = r \geq d_G(v_r) \geq \delta \geq k$. □

1.5. Automorphism of a Simple Graph

Definition 1.5.1 An *automorphism* of a graph G is an isomorphism of G onto itself. We recall that two simple graphs G and H are isomorphic if, and only if, there exists a bijection $\phi : V(G) \rightarrow V(H)$ such that uv is an edge of G if, and only if, $\phi(u)\phi(v)$ is an edge of H. In this case ϕ is called an isomorphism of G onto H.

We prove in our next theorem that the set $\Gamma(G)$ of automorphisms of G is a group.

Theorem 1.5.2 *The set $\Gamma(G)$ of all automorphisms of a simple graph G is a group with respect to the composition $\circ$ of mappings as the group operation.*

Proof We shall verify that the four axioms of a group are satisfied by the pair $(\Gamma(G), \circ)$.

(i) Let ϕ_1 and ϕ_2 be bijections on $V(G)$ preserving adjacency and nonadjacency. Clearly the mapping $\phi_1 \circ \phi_2$ is a bijection on $V(G)$. If u and v are adjacent in G, then $\phi_2(u)$ and $\phi_2(v)$ are adjacent in G. But $(\phi_1 \circ \phi_2)(u) = \phi_1(\phi_2(u))$ and $(\phi_1 \circ \phi_2)(v) = \phi_1(\phi_2(v))$. Hence $(\phi_1 \circ \phi_2)(u)$ and $(\phi_1 \circ \phi_2)(v)$ are adjacent in G; that is, $\phi_1 \circ \phi_2$ preserves adjacency. A similar argument shows that $\phi_1 \circ \phi_2$ preserves nonadjacency. Thus $\phi_1 \circ \phi_2$ is an automorphism of G.

(ii) It is a well-known result that the composition of mappings of a set onto itself is associative.

(iii) The identity mapping I of $V(G)$ onto itself is an automorphism of G and it satisfies the condition $\phi \circ I = I \circ \phi = \phi$ for every $\phi \in \Gamma(G)$. Hence I is the identity element of $\Gamma(G)$.

(iv) Finally, if ϕ is an automorphism of G, the inverse mapping ϕ^{-1} is also an automorphism of G (see Section 1.1). □

Theorem 1.5.3 *For any simple graph G, $\Gamma(G) = \Gamma(G^c)$.*

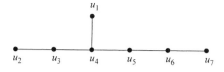

FIGURE 1.21. An identity graph

Proof Since $V(G^c) = V(G)$, every bijection on $V(G)$ is also a bijection on $V(G^c)$. As an automorphism of G preserves adjacency and nonadjacency of vertices of G, it also preserves adjacency and nonadjacency of vertices of G^c. Hence every element of $\Gamma(G)$ is also an element of $\Gamma(G^c)$ and vice versa. □

Exercise 5.1 Show that the automorphism group of K_n (or K_n^c) is isomorphic to the symmetric group S_n of degree n.

In contrast to the complete graphs for which the automorphism group consists of every bijection of the vertex set, there are graphs whose automorphism groups consist of just the identity permutation. Such graphs are called *identity graphs*.

Example 1.5.4 The graph G shown in Figure 1.21 is an identity graph.

Solution Let γ be an automorphism of G. Then γ preserves degrees; that is, $d(v) = d(\gamma(v))$ for all $v \in V(G)$ (see Exercise 3.1). Since u_4 is the only vertex of degree 3 in G, $\gamma(u_4) = u_4$. Now, u_1, u_2, and u_7 are the vertices of degree 1 in G. Hence $\gamma(u_1) \in \{u_1, u_2, u_7\}$. Also, since u_1 is adjacent to u_4, $\gamma(u_1)$ is adjacent to $\gamma(u_4) = u_4$. Hence the only possibility is $\gamma(u_1) = u_1$. Now, u_3, u_5, and u_6 are the vertices of degree 2 in G. Hence $\gamma(u_3) \in \{u_3, u_5, u_6\}$. Also, as u_3 is adjacent to u_4, $\gamma(u_3)$ is adjacent to $\gamma(u_4) = u_4$. Hence $\gamma(u_3)$ is u_3 or u_5. Again, $\gamma(u_3)$ is adjacent to a vertex of degree 1. This forces $\gamma(u_3) \neq u_5$ since u_5 is not adjacent to a vertex of degree 1. Consequently, $\gamma(u_3) = u_3$. This again forces $\gamma(u_2) = u_2$. Having proved $\gamma(u_2) = u_2$, we must have $\gamma(u_7) = u_7$ on degree consideration. Using similar arguments, one easily proves that $\gamma(u_5) = u_5$ and $\gamma(u_6) = u_6$. Consequently, $\gamma = I$, which implies that $\Gamma(G) = \{I\}$. □

Exercise 5.2 Let G be a simple connected graph with n vertices such that $\Gamma(G) = S_n$. Show that G is the complete graph K_n.

Exercise 5.3 For $n > 1$, give (a) simple connected graph $G \neq K_n$ with $\Gamma(G) \cong S_n$ and (b) a simple disconnected graph with $G \neq K_n$ with $\Gamma(G) \simeq S_n$.

Exercise 5.4 Find the automorphism groups of the following graphs:

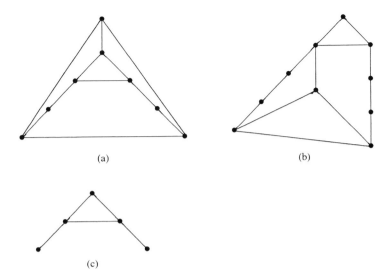

(a)   (b)

(c)

Exercise 5.5 Let G be a simple graph and $\gamma \in \Gamma(G)$. Prove that $\gamma\{N(v)\} = N(\gamma(v))$ and $\gamma\{N[v]\} = N[\gamma(v)]$ for every $v \in V$.

1.6.   Line Graphs

Let G be a loopless graph. We construct a graph $L(G)$ in the following way:

The vertex set of $L(G)$ is in 1–1 correspondence with the edge set of G and two vertices of $L(G)$ are joined by an edge if, and only if, the corresponding edges of G are adjacent in G. The graph $L(G)$ (which is always a simple graph) is called the *line graph* or the *edge graph* of G.

Figure 1.22 shows a graph and its line graph in which v_i of $L(G)$ corresponds to e_i of G for each i. Isolated vertices of G do not have any bearing on $L(G)$ and hence we assume in this section that G has no isolated vertices. We also assume that G has no loops. (See Exercise 6.3.)

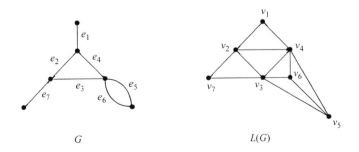

G $L(G)$

FIGURE 1.22. A graph G and its line graph $L(G)$

Some simple properties of the line graph $L(G)$ of a graph G follow:

1. G is connected if, and only if, $L(G)$ is connected.
2. If H is a subgraph of G, then $L(H)$ is a subgraph of $L(G)$.
3. The edges incident at a vertex of G give rise to a maximal complete subgraph of $L(G)$.
4. If e is an edge of G joining u and v, then the degree of e in $L(G)$ is the same as the number of edges of G adjacent to e in G. This number is the same as $d_G(u) + d_G(v) - 2$. Hence $d_{L(G)}(e) = d_G(u) + d_G(v) - 2$.
5. Finally,

$$\sum_{e \in V(L(G))} d_{L(G)}(e) = \sum_{uv \in E(G)} (d_G(u) + d_G(v) - 2)$$

$$= \left[\sum_{u \in V(G)} d_G(u)^2 \right] - 2m(G) \quad \text{(since } uv \text{ belongs to the stars at } u \text{ and } v; \text{ see Definitions 1.1.12)}$$

$$= \left[\sum_{i=1}^{n} d_i^2 \right] - 2m.$$

where $(d_1, d_2, \ldots, d_n)$ is the degree sequence of G, and $m = m(G)$. By Euler's theorem (Theorem 1.3.4), it follows that

$$m(L(G)) = \frac{1}{2} \left[\sum_{i=1}^{n} d_i^2 \right] - m.$$

Exercise 6.1 Show that the line graph of the star $K_{1,n}$ is the complete graph K_n.

Exercise 6.2 Show that $L(C_n) \cong C_n, n \geq 3$.

Theorem 1.6.1 *The line graph of a simple graph G is a path if, and only if, G is a path.*

Proof Let G be the path P_n on n vertices. Then clearly $L(G)$ is the path P_{n-1} on $n - 1$ vertices.

Conversely, let $L(G)$ be a path. Then no vertex of G can have degree greater than 2 because, if G has a vertex v of degree greater than 2, the edges incident to v would form a complete subgraph of $L(G)$ with at least three vertices. Hence G must be either a cycle or a path. But G cannot be a cycle, because the line graph of a cycle is a cycle. ☐

Exercise 6.3 Give an example of a graph G to show that the relation $d_{L(G)}(e) = d_G(u) + d_G(v) - 2$ may not be valid if G has a loop.

As shown in Exercise 6.2, $C_n \cong L(C_n)$; in fact, C_n is the only simple graph with this property as Exercise 6.4 shows.

Exercise 6.4 Prove that a simple connected graph G is isomorphic to its line graph if, and only if, it is a cycle.

Exercise 6.5 Disprove: H is a spanning subgraph of G implies that $L(H)$ is a spanning subgraph of $L(G)$.

Theorem 1.6.2   *If the simple graphs G_1 and G_2 are isomorphic, then $L(G_1)$ and $L(G_2)$ are isomorphic.*

Proof   Let (ϕ, θ) be an isomorphism of G_1 onto G_2. Then θ is a bijection of $E(G_1)$ onto $E(G_2)$. We show that θ is an isomorphism of $L(G_1)$ to $L(G_2)$. We prove this by showing that θ preserves adjacency and nonadjacency. Let e_i and e_j be two adjacent vertices of $L(G_1)$. Then there exists a vertex v of G_1 incident with both e_i and e_j and so $\phi(v)$ is a vertex incident with both $\theta(e_i)$ and $\theta(e_j)$. Hence $\theta(e_i)$ and $\theta(e_j)$ are adjacent vertices in $L(G_2)$.

Now, let $\theta(e_i)$ and $\theta(e_j)$ be adjacent vertices in $L(G_2)$. This means that they are adjacent edges in G_2 and hence there exists a vertex v' of G_2 incident to both $\theta(e_i)$ and $\theta(e_j)$. Then $\phi^{-1}(v')$ is a vertex of G_1 incident to both e_i and e_j so that e_i and e_j are adjacent vertices of $L(G_1)$.

Thus, e_i and e_j are adjacent vertices of $L(G_1)$ if, and only if, $\theta(e_i)$ and $\theta(e_j)$ are adjacent vertices of $L(G_2)$. Hence θ is an isomorphism of $L(G)_1$ onto $L(G_2)$. (Recall that a line graph is always a simple graph.)   □

Remark 1.6.3   The converse of Theorem 1.6.2 is not true. Consider the graphs $K_{1,3}$ and K_3. Their line graphs are K_3. But $K_{1,3}$ is not isomorphic to K_3 since there is a vertex of degree 3 in $K_{1,3}$ whereas there is no such vertex in K_3.

Theorem 1.6.4 shows that the above two graphs are the only two exceptional simple graphs of this type.

Theorem 1.6.4* (H. Whitney)   *Let G and G' be simple connected graphs with isomorphic line graphs. Then G and G' are isomorphic unless one of them is $K_{1,3}$ and the other is K_3.*

Proof   First suppose that $n(G)$ and $n(G')$ are less than or equal to 4. A necessary condition for $L(G)$ and $L(G')$ to be isomorphic is that $m(G) = m(G')$. The only nonisomorphic connected graphs on at most four vertices are those shown in Figure 1.23.

In Figure 1.23, graphs G_4, G_5, and G_6 are the three graphs having three edges each. We have already seen that G_4 and G_6 have isomorphic line graphs, namely K_3. The line graph of G_5 is a path of length 2, and hence $L(G_5)$ cannot be isomorphic to $L(G_4)$ or $L(G_6)$. Further, G_7 and G_8 are the only two graphs in the list having four edges each.

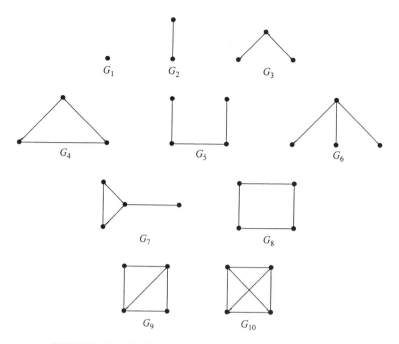

FIGURE 1.23. Nonisomorphic graphs on four vertices or less

Now, $L(G_8) \cong G_8$, and $L(G_7)$ is isomorphic to G_9. Thus, the line graphs of G_7 and G_8 are not isomorphic. No two of the remaining grpahs have the same number of edges. Hence, the only nonisomorphic graphs with at most four vertices having isomorphic line graphs are G_4 and G_6.

We now suppose that either G or G', say G, has at least five vertices and that $L(G)$ and $L(G')$ are isomorphic under an isomorphism ϕ_1. ϕ_1 is a bijection from the edge set of G onto the edge set of G'.

We now prove that ϕ_1 transforms an induced $K_{1,3}$ subgraph of G onto a $K_{1,3}$ subgraph of G'. Let $e_1 = uv_1$, $e_2 = uv_2$ and $e_3 = uv_3$ be the edges of an induced $K_{1,3}$ subgraph of G. As G has at least five vertices and is connected, there exists an edge e adjacent to only one or all three of edges e_1, e_2, and e_3, as illustrated in Figure 1.24.

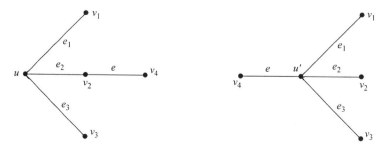

FIGURE 1.24. Graphs with five vertices and edge e adjacent to one or all three other edges

Now, $\phi_1(e_1)$, $\phi_1(e_2)$, and $\phi_1(e_3)$ form either a $K_{1,3}$ subgraph or a triangle in G'. If $\phi_1(e_1)$, $\phi_1(e_2)$, and $\phi_1(e_3)$ form a triangle in G', $\phi_1(e)$ can be adjacent to precisely two of $\phi_1(e_1)$, $\phi_1(e_2)$, and $\phi_1(e_3)$ (since $L(G')$ is simple), whereas $\phi_1(e)$ must be adjacent to only one or all the three. This contradiction shows that $\{\phi_1(e_1), \phi_1(e_2), \phi_1(e_3)\}$ is not a triangle in G' and therefore forms a $K_{1,3}$ in G'.

It is clear that a similar result holds as well for ϕ_1^{-1}, since it is an isomorphism of $L(G')$ onto $L(G)$.

Let $S(u)$ denote the star subgraph of G formed by the edges of G incident at a vertex u of G. We shall prove that ϕ_1 maps $S(u)$ onto the unique star subgraph $S(u')$ of G'.

(i) First suppose that the degree of u is at least 2. Let f_1 and f_2 be any two edges incident at u. The edges $\phi_1(f_1)$ and $\phi_1(f_2)$ of G' have an end vertex u' in common. If f is any other edge of G incident with u, then $\phi_1(f)$ is incident with u', and conversely, for every edge f' of G' incident with u', $\phi_1^{-1}(f')$ is incident with u. Thus $S(u)$ in G is mapped to $S(u')$ in G'.

(ii) Let the degree of u be 1 and $e = uv$ be the unique edge incident with u. As G is connected and $n(G) \geq 5$, degree of v must be at least 2 in G, and therefore, by (i), $S(v)$ is mapped to a star $S(v')$ in G'. Also $\phi_1(uv) = u'v'$ for some $u' \in V(G')$. Now, if the degree of u' is greater than 1, by paragraph (i), the star at u' in G' is transformed by ϕ_1^{-1} either to the star at u in G or to the star at v in G. But as the star at v in G is mapped to the star at v' in G' by ϕ_1, ϕ_1^{-1} should map the star at u' to the star at u only. As ϕ_1^{-1} is 1–1, this means that $\deg u \geq 2$, a contradiction. Therefore, $\deg u' = 1$, and so $S(u)$ in G is mapped onto $S(u')$ in G'.

We now define $\phi : V(G) \to V(G')$ by setting $\phi(u) = u'$ if $\phi_1(S(u)) = S(u')$. Since $S(u) = S(v)$ only when $u = v$ ($G \neq K_2$, $G' \neq K_2$), ϕ is 1–1. ϕ is also onto since, for v' in G', $\phi_1^{-1}(S(v')) = S(v)$ for some $v \in V(G)$, and by the definition of ϕ, $\phi(v) = v'$. Finally, if uv is an edge of G, then $\phi_1(uv)$ belongs to both $S(u')$ and $S(v')$, where $\phi_1(S(u)) = S(u')$ and $\phi_1(S(v)) = S(v')$. This means that $u'v'$ is an edge of G'. But $u' = \phi(u)$ and $v' = \phi(v)$. Consequently, $\phi(u)\phi(v)$ is an edge of G'. If u and v are nonadjacent in G, $\phi(u)\phi(v)$ must be nonadjacent in G'. Otherwise, $\phi(u)\phi(v)$ belongs to both $S(\phi(u))$ and $S(\phi(v))$, and hence $\phi_1^{-1}(\phi(u)\phi(v)) = uv \in E(G)$, a contradiction. Thus G and G' are isomorphic under ϕ. □

Definition 1.6.5 A graph H is called a *forbidden subgraph* for a property P of graphs if it satisfies the following condition: If a graph G has property P, then G cannot contain an induced subgraph isomorphic to H.

Beineke [10] obtained a forbidden subgraph criterion for a graph to be a line graph. In fact, he showed that a graph G is a line graph if, and only if, the nine graphs of Figure 1.25 are forbidden subgraphs for G. However, for the sake of later reference, we prove the following result.

Theorem 1.6.6 *If G is a line graph, then $K_{1,3}$ is a forbidden subgraph of G.*

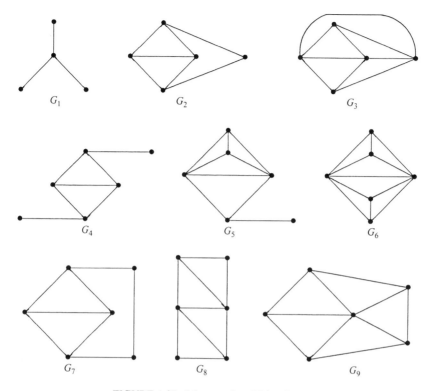

FIGURE 1.25. Nine graphs of Bieneke [10]

Proof   Suppose that G is the line graph of graph H, and that G contains $K_{1,3}$ as an induced subgraph. If v is the vertex of degree 3 in $K_{1,3}$ and v_1, v_2, and v_3 are the neighbors of v of this $K_{1,3}$, then the edge e corresponding to v in H is adjacent to the three edges e_1, e_2, and e_3 corresponding to the vertices v_1, v_2, and v_3. Hence, one of the end vertices of e must be the end vertex of at least two of e_1, e_2, and e_3 in H, and hence v together with two of v_1, v_2, and v_3 form a triangle in G. This means that the $K_{1,3}$ subgraph of G considered here is not an induced subgraph of G, a contradiction.   □

1.7.   Operations on Graphs

In mathematics, one always tries to get new structures from given ones. This also applies to the realm of graphs, where one can generate many new graphs from a given set of graphs. In this section we consider some of the methods of generating new graphs from a given pair of graphs.

Let $G_1 = (V_1, E_1)$ and $G_2 = (V_2, E_2)$ be two simple graphs.

Definition 1.7.1   *Union of two graphs*: The graph $G = (V, E)$, where $V = V_1 \cup V_2$ and $E = E_1 \cup E_2$ is called the union of G_1 and G_2 and is denoted by $G_1 \cup G_2$.

When G_1 and G_2 are vertex disjoint, $G_1 \cup G_2$ is denoted by $G_1 + G_2$ and is called the *sum* of the graphs G_1 and G_2.

The finite union of graphs is defined by means of associativity; in particular, if $G_1, G_2, \ldots, G_r$ are pairwise vertex disjoint graphs, each of which is isomorphic to G, then $G_1 + G_2 + \ldots + G_r$ is denoted by rG.

Definition 1.7.2   *Intersection of two graphs*: If $V_1 \cap V_2 \neq \phi$, the graph $G = (V, E)$, where $V = V_1 \cap V_2$ and $E = E_1 \cap E_2$, is the *intersection* of G_1 and G_2 and is written as $G_1 \cap G_2$.

Definition 1.7.3   *Join of two graphs*: Let G_1 and G_2 be *vertex-disjoint* graphs. Then the *join*, $G_1 \vee G_2$, of G_1 and G_2 is the supergraph of $G_1 + G_2$ in which each vertex of G_1 is adjacent to every vertex of G_2.

Figure 1.26 illustrates the graph $G_1 \vee G_2$. If $G_1 = K_1$ and $G_2 = C_n$, then $G_1 \vee G_2$ is called the *wheel* W_n. W_5 is shown in Figure 1.27.

It is worthwhile to note that $K_{m,n} = K_m^c \vee K_n^c$ and $K_n = K_1 \vee K_{n-1}$.

It follows from the above definitions that

1. $n(G_1 \cup G_2) = n(G_1) + n(G_2) - n(G_1 \cap G_2)$,
    $m(G_1 \cup G_2) = m(G_1) + m(G_2) - m(G_1 \cap G_2)$.
2. $n(G_1 + G_2) = n(G_1) + n(G_2)$, $m(G_1 + G_2) = m(G_1) + m(G_2)$.
3. $n(G_1 \vee G_2) = n(G_1) + n(G_2)$, $m(G_1 \vee G_2) = m(G_1) + m(G_2) + n(G_1)n(G_2)$.

Definition 1.7.4   The *cartesian product* $G_1 \times G_2$ of two graph G_1 and G_2 is the simple graph with $V_1 \times V_2$ as its vertex set and two vertices (u_1, v_1) and (u_2, v_2) are adjacent in $G_1 \times G_2$ if, and only if, either $u_1 = u_2$ and v_1 is adjacent to v_2 in G_2, or u_1 is adjacent to u_2 in G_1 and $v_1 = v_2$.

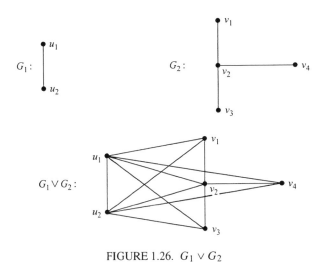

FIGURE 1.26. $G_1 \vee G_2$

FIGURE 1.27. Wheel W_5

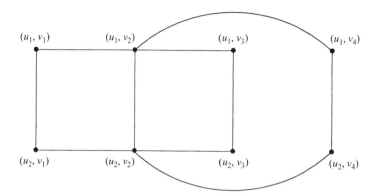

FIGURE 1.28. $G_1 \times G_2$

It is easy to see that $n(G_1 \times G_2) = n(G_1)n(G_2)$ and $m(G_1 \times G_2) = n(G_1)m(G_2) + m(G_1)n(G_2)$.

Example 1.7.5 If G_1 and G_2 are graphs of Figure 1.26, then $G_1 \times G_2$ is the graph of Figure 1.28.

Definition 1.7.6 Composition or lexicographic product: Let G_1 and G_2 be two simple graphs. Then the simple graph with $V_1 \times V_2$ as the vertex set and in which the vertices $u = (u_1, u_2)$, $v = (v_1, v_2)$ are adjacent if either u_1 is adjacent to v_1 or $u_1 = v_1$, and u_2 is adjacent to v_2 is defined as the *composition* or *lexicographic product* of G_1 and G_2 and is denoted by $G_1[G_2]$.

Example 1.7.7 If G_1 and G_2 are graphs of Figure 1.26, then $G_1[G_2]$ and $G_2[G_1]$ are shown in Figures 1.29 and 1.30, respectively.

Exercise 7.1 Show by means of an example that $G_1[G_2]$ need not be isomorphic to $G_2[G_1]$.

Exercise 7.2 Prove that $G_1 \times G_2 \cong G_2 \times G_1$.

We now define another graph product.

Definition 1.7.8 *Normal product or strong product:* The *normal product* $G_1 \circ G_2$ of two simple graphs G_1 and G_2 is the graph with $V(G_1 \circ G_2) = V_1 \times V_2$ wherein (u_1, u_2) and (v_1, v_2) are adjacent in $G_1 \circ G_2$ if, and only if, either

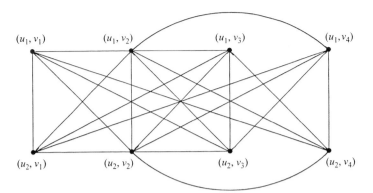

FIGURE 1.29. $G_1[G_2]$

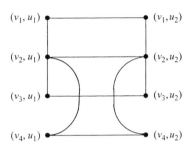

FIGURE 1.30. $G_2[G_1]$

(i) $u_1 = v_1$ and u_2 is adjacent to v_2, or
(ii) u_1 is adjacent to v_1 and $u_2 = v_2$, or
(iii) u_1 is adjacent to v_1 and u_2 is adjacent to v_2.

Definition 1.7.9 *Tensor product or Kronecker product*: The *tensor product* $G_1 \otimes G_2$ of two simple graphs G_1 and G_2 is the graph with $V(G_1 \otimes G_2) = V_1 \times V_2$ and wherein (u_1, u_2) and (v_1, v_2) are adjacent in $G_1 \otimes G_2$ if, and only if, u_1 is adjacent to v_1 in G_1 and u_2 is adjacent to v_2 in G_2.
  We note that $G_1 \circ G_2 = (G_1 \times G_2) \cup (G_1 \otimes G_2)$.

Example 1.7.10 If G_1 and G_2 are graphs of Figure 1.26, then $G_1 \circ G_2$ and $G_1 \otimes G_2$ are shown in Figure 1.31.

Exercise 7.3 Prove that $G_1 \circ G_2$ is isomorphic to $G_2 \circ G_1$.

Exercise 7.4 Prove that

(a) $n(G_1[G_2]) = n(G_2[G_1]) = n(G_1 \circ G_2) = n(G_1)n(G_2)$
(b) $m(G_1[G_2]) = n(G_1)m(G_2) + n(G_2)^2 m(G_1)$
(c) $m(G_1 \circ G_2) = n(G_1)m(G_2) + m(G_1)n(G_2) + 2m(G_1)m(G_2)$
(d) $m(G_1 \otimes G_2) = 2m(G_1)m(G_2)$.

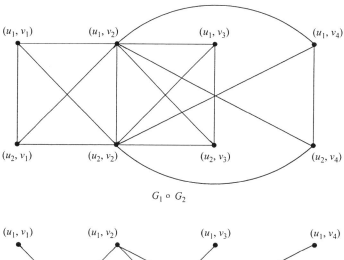

$$G_1 \circ G_2$$

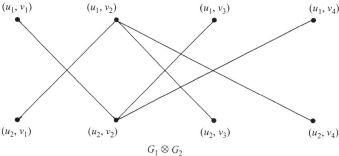

$$G_1 \otimes G_2$$

FIGURE 1.31. $G_1 \circ G_2$ and $G_1 \otimes G_2$

We now introduce the powers of a simple graph G.

Definition 1.7.11   The *kth power* G^k of G has $V(G^k) = V(G)$, where u and v are adjacent in G^k whenever $d_G(u, v) \leq k$.

Two graphs H_1 and H_2 and their squares H_1^2 and H_2^2 are displayed in Figure 1.32. By definition of G^k, G is a *spanning subgraph* of G^k, $k \geq 1$.

Exercise 7.5   Show that if G is a connected graph with at least two edges, then each edge of G^2 belongs to a triangle.

Exercise 7.6   If $d_G(u, v) = m$, determine $d_{G^k}(u, v)$.

1.8.   An Application to Chemistry

The earliest application of graph theory to chemistry was found by Cayley [23] in 1857 when he enumerated the number of rooted trees with a given number of vertices. A chemical molecule can be represented by a graph by taking each atom of the molecule as a vertex of the graph and making the edges of the graph represent atomic bonds between the end atoms. Thus the degree of each vertex of

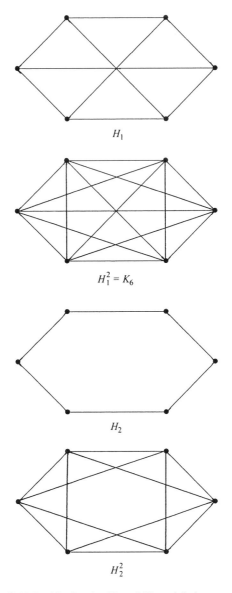

FIGURE 1.32. Graphs H_1 and H_2 and their squares

such a graph gives the valance of the corresponding atom. Molecules that have the same chemical formula but have different structural properties are called isomers. In terms of graphs, isomers are two nonisomorphic graphs having the same degree sequence.

The two graphs (actually trees; see chapter IV) of Figure 1.33 represent two isomers of the molecule C_3H_7OH (propanol). The graph of Figure 1.34 represents

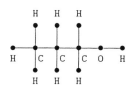
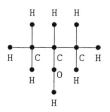

FIGURE 1.33. Two isomers of C_3H_7OH

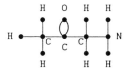

FIGURE 1.34. Graph of C_3H_7NO

aminoacetone C_3H_7NO. This has a multiple bond represented by a pair of multiple edges between C and O.

The paraffins have the molecular formula C_kH_{2k+2}. They have $3k + 2$ atoms (vertices) of which k are carbon atoms and the remaining $2k + 2$ are hydrogen atoms. They all have $3k + 1$ bonds (edges). Cayley used enumeration techniques of graph theory (see reference [65]) to count the number of isomers of C_kH_{2k+2}. His formula shows that for the paraffin $C_{13}H_{28}$, there are 802 different isomers.

1.9. Miscellaneous Exercises

9.1   Show that the sequences (a) (7, 6, 5, 4, 3, 3, 2) and (b) (6, 6, 5, 4, 3, 3, 1) are not graphical.

9.2   Give an example of a degree sequence that is realizable as the degree sequence by only a disconnected graph.

9.3   Show that for a simple bipartite graph, $m \leq \frac{n^2}{4}$.

9.4   If δ and Δ are, respectively, the minimum and maximum of the degrees of a graph G, then show that $\delta \leq \frac{2m}{n} \leq \Delta$.

9.5   For every $n \geq 3$, construct a 3-regular simple graph with $2n$ vertices containing no triangles.

9.6   If a bipartite graph $G(X, Y)$ is regular, show that $|X| = |Y|$.

9.7   Show that in a simple graph, a closed walk of odd length contains a cycle.

9.8   Give an example of a disconnected simple graph having ω components, n vertices, and $\frac{(n-\omega)(n-\omega+1)}{2}$ edges.

9.9   Prove or disprove: If H is a subgraph of G, then

(a) $\delta(H) \le \delta(G)$

(b) $\Delta(H) \le \Delta(G)$.

9.10   If $m > \frac{(n-1)(n-2)}{2}$, then show that G is connected.

9.11   If $\delta \ge 2$, then show that G contains a cycle.

9.12   If $\delta(G) \ge 3k - 1$ for a graph G, prove that G contains k edge-disjoint cycles.

9.13   If a simple graph G has two pendant vertices, prove that G^c has at most two pendant vertices. Give an example of a graph G such that both G and G^c have exactly two pendant vertices.

9.14   Show that $\Gamma(C_n)$ is D_n, the dihedral group of order $2n$.

9.15   Determine $\Gamma(K_{p,q})$, $p \ne q$.

9.16   Determine the order of the automorphism group of
       (a) $K_4 - e$; (b) P_n.

9.17   Show that the line graph of K_n is regular of degree $2n - 4$. Draw the line graph of K_4.

9.18   Show that the line graph of $K_{p,q}$ is regular of degree $p + q - 2$. Draw the line graph of $K_{2,3}$.

9.19   Show that a connected graph G is complete bipartite if, and only if, no induced subgraph of G is a K_3 or P_4.

9.20   If G is connected and diam$(G) \ge 3$, where diam$(G) = \max\{d(u, v) : u, v \in V(G)\}$, then show that G^c is connected and diam$(G^c) \le 3$.

9.21   Show that the complement of a simple connected graph G is connected if, and only if, G contains no spanning complete bipartite subgraph.

Notes

Graph theory, which had arisen out of puzzles solved for the sake of curiosity, has now grown into a major discipline in mathematics with problems permeating into almost all subjects—physics, chemistry, engineering, psychology, computer science, and what not! It is customary to assume that Graph theory originated with Leonhard Euler (1701–1783), who formulated the first few theorems in the subject. The subject, which has been lying almost dormant for more than hundred years since the time of Euler, suddenly started exploding by the turn of twentieth century, and today it has branched off in various directions—coloring problems, Ramsey theory, hypergraph theory, Eulerian and Hamiltonian graphs, decomposition and factorization theory, directed graphs, just to name a few. Some of the standard texts in graph theory are the references [9], [13], [19], [25], [29], [53], [61], and [103]. A good account of enumeration theory of graphs is given in reference [65]. Further, a comprehensive account of applications of graph theory to chemistry is given in references [113] and [114].

Theorem 1.6.4 is due to H. Whitney [121], and the proof given in this chapter is due to Jüng [74].

II
DIRECTED GRAPHS

2.0. Introduction

Directed graphs arise in a natural way in many applications of graph theory. The street map of a city, abstract representation of computer programs, and network flows can be represented only by directed graphs rather than by graphs. Directed graphs also are used in the study of sequential machines and system analysis in control theory.

2.1. Basic Concepts

Definition 2.1.1 A *directed graph* D is an ordered triple $(V(D), A(D), I_D)$, where $V(D)$ is a nonempty set called the set of *vertices* of D; $A(D)$ is a set disjoint from $V(D)$, called the set of *arcs* of D; and I_D is an *incidence map* that associates with each arc of D an ordered pair of vertices of D. If a is an arc of D, and $I_D(a) = (u, v)$, u is called the *tail* of a, and v is the *head* of a. The arc a is said to join v with u. u and v are called the ends of a. A directed graph is also called a *digraph*.

With each digraph D, we can associate a graph G (written $G(D)$ when reference to D is needed) on the same vertex set as follows: corresponding to each arc of D, there is an edge of G with the same ends. This graph G is called the *underlying graph* of the digraph D. Thus every digraph D defines a unique (up to isomorphism) graph G. Conversely, given any graph G, we can obtain a digraph

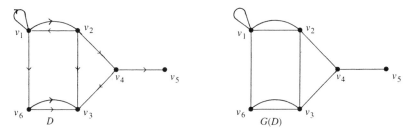

FIGURE 2.1. Digraph D and its underlying graph G

from G by specifying for each edge of G an order of its ends. Such a specification is called an *orientation* of G.

Just as with graphs, digraphs have diagrammatic representation. A digraph is represented by a diagram of its underlying graph together with arrows on its edges, the arrow pointing toward the head of the corresponding arc. A digraph and its underlying graph are shown in Figure 2.1.

Many of the concepts and terminology for graphs are also valid for digraphs. However, there are many concepts of digraphs involving the notion of orientation that apply only to digraphs.

Definition 2.1.2   If $a = (u, v)$ is an arc of D, a is said to be *incident out of* u and *incident into* v. v is called an *outneighbor* of u, and u is called an *inneighbor* of v. $N_D^+(u)$ denotes the set of outneighbors of u in D. Similarly, $N_D^-(u)$ denotes the set of inneighbors of u in D. When no explicit reference to D is needed, we denote these sets by $N^+(u)$ and $N^-(u)$, respectively. An arc a is *incident with* u if either it is incident into or incident out of u. An arc having the same ends is called a *loop* of D. The number of arcs incident out of a vertex v is the *outdegree of* v and it is denoted by $d_D^+(v)$ or $d^+(v)$. The number of arcs incident into v is its *indegree* and is denoted by $d_D^-(v)$ or $d^-(v)$.

For the digraph D of Figure 2.1, we have $d^+(v_1) = 3, d^+(v_2) = 3, d^+(v_3) = 0, d^+(v_4) = 2, d^+(v_5) = 0, d^+(v_6) = 2, d^-(v_1) = 2, d^-(v_2) = 1, d^-(v_3) = 4, d^-(v_4) = 1, d^-(v_5) = 1$, and $d^-(v_6) = 1$. (The loop at v_1 contributes 1 each to $d^+(v_1)$ and $d^-(v_1)$.)

The *degree* $d(v)$ of a vertex v of a digraph D is the degree of v in $G(D)$. Thus $d(v) = d^+(v) + d^-(v)$. As each arc of a digraph contributes 1 to the sum of the outdegrees and 1 to the sum of indegrees, we have

$$\sum_{v \in V(D)} d^+(v) = \sum_{v \in V(D)} d^-(v) = m(D),$$

where $m(D)$ is the number of arcs of D.

A vertex of D is *isolated* if its degree is 0; it is *pendant* if its degree is 1. Thus for a pendant vertex v, either $d^+(v) = 1$ and $d^-(v) = 0$, or $d^+(v) = 0$ and $d^-(v) = 1$.

FIGURE 2.2. A strong digraph (left) and a symmetric digraph (right)

Definitions 2.1.3

1. A digraph D' is a *subdigraph* of a digraph D if $V(D') \subseteq V(D)$, $A(D') \subseteq A(D)$, and $I_{D'}$ is the restriction of I_D to $A(D')$.

2. A *directed walk* joining the vertex v_0 to the vertex v_k in D is an alternating sequence $w = v_0 a_1 v_1 a_2 v_2 \ldots a_k v_k$, $1 \leq i \leq k$, with a_i incident out of v_{i-1} and incident into v_i. *Directed trails, directed paths, directed cycles*, and *induced subdigraphs* are defined analogously as for graphs.

3. A vertex v is *reachable* from a vertex u of D if there is a directed path in D from u to v.

4. Two vertices of D are *diconnected* if each is reachable from the other. Clearly, diconnection is an equivalence relation on the vertex set of D and if the equivalence classes are $V_1, V_2, \ldots, V_\omega$, the subdigraphs of D induced by $V_1, V_2, \ldots, V_\omega$ are called the *dicomponents* of D.

5. A digraph is *diconnected* (also called *strongly connected*) if it has exactly one dicomponent. A diconnected digraph is also called a *strong digraph*.

6. A digraph is *strict* if its underlying graph is simple. A digraph D is *symmetric* if whenever (u, v) is an arc of D, then (v, u) is also an arc of D. (See Figure 2.2.)

Exercise 1.1 How many orientations does a simple graph on n vertices have?

Exercise 1.2 Let D be a digraph with no directed cycle. Prove that there exists a vertex whose indegree is 0. Deduce that there is an ordering $v_1, v_2, \ldots, v_n$ of V such that, for $1 \leq i \leq n$, every arc of D with terminal vertex v_i has its initial vertex in $\{v_1, v_2, \ldots, v_{i-1}\}$.

2.2. Tournaments

A digraph D is a *tournament* if its underlying graph is a complete graph. Thus, in a tournament, for every pair of distinct vertices u and v, either (u, v) or (v, u), but not both, is an arc of D. Figures 2.3 (a) and (b) display all tournaments on three and four vertices, respectively.

The word "tournament" derives its name from the usual round-robin tournament. Suppose there are n players in a tournament and that every player is to play against every other player. The results of such a tournament can be represented by a tournament on n vertices where the vertices represent the n players and an arc (u, v) represents the victory of the player u over the player v.

Suppose the players of a tournament have to be ranked. The corresponding digraph could be used for such a ranking. This is because any tournament contains

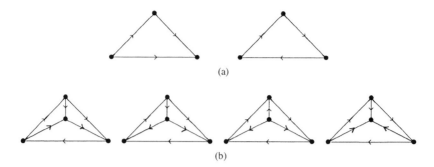

FIGURE 2.3. Tournaments on (a) three and (b) four vertices

a *directed Hamilton path*, that is, a spanning directed path. One could rank the players as per the sequence of this path so that each player defeats his successor. We now prove the existence of a directed Hamilton path in any tournament.

Theorem 2.2.1 (Rédei [107])   *Every tournament contains a directed Hamilton path.*

Proof   (By induction on the number of vertices n of the tournament.) The result can be directly verified for all tournaments up to three vertices. Hence suppose that the result is true for all tournaments on $n \geq 3$ vertices. Let T be a tournament on $(n + 1)$ vertices $v_1, v_2, \ldots, v_{n+1}$. Now delete v_{n+1} from T. The resulting digraph T' is a tournament on n vertices and hence by the induction hypothesis contains a directed Hamilton path. Assume that the Hamilton path is $v_1 v_2 \ldots v_n$, relabeling the vertices, if necessary.

If the arc joining v_1 and v_{n+1} has v_{n+1} as its tail, then $v_{n+1} v_1 v_2 \cdots v_n$ is a directed Hamilton path in T, and the result stands proved. (See Figure 2.4 (a).)

If the arc joining v_n and v_{n+1} is directed from v_n to v_{n+1}, then $v_1 v_2 \cdots v_n v_{n+1}$ is a directed Hamilton path in T. (See Figure 2.4 (b).)

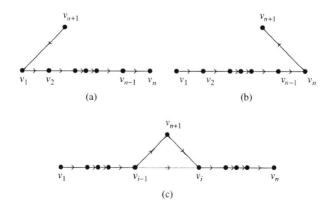

FIGURE 2.4. Graphs for proof of Theorem 2.2.1

Now suppose that none of (v_{n+1}, v_1) and (v_n, v_{n+1}) is an arc of T. Hence (v_1, v_{n+1}) and (v_{n+1}, v_n) are arcs of T—the first incident into v_{n+1} and the second out of v_{n+1}. Thus, as we pass on from v_1 to v_n we encounter a reversal of the orientation of edges incident with v_{n+1}. Let v_i, $2 \le i \le n$, be the first vertex where this reversal takes place, so that (v_{i-1}, v_{n+1}) and (v_{n+1}, v_i) are arcs of T. Then $v_1 v_2 \ldots v_{i-1} v_{n+1} v_i v_{i+1} \ldots v_n$ is a directed Hamilton path of T. (See Figure 2.4 (c).) $\square$

Theorem 2.2.2 (Moon [89, 91])   *Every vertex of a diconnected tournament T with $n \ge 3$ is contained in a directed k-cycle, $3 \le k \le n$. (T is then said to be vertex pancyclic.)*

Proof   Let T be a diconnected tournament with $n \ge 3$ and u be a vertex of T. Let $S = N^+(u)$, the set of all outneighbors of u in T, and $S' = N^-(u)$, the set of all inneighbors of u in T. As T is diconnected, none of S and S' is empty. If $[S, S']$ denotes the set of all arcs of T having their tails in S and heads in S', then $[S, S']$ is also nonempty for the same reason. If (v, w) is an arc of $[S, S']$, then (u, v, w, u) is a directed 3-cycle in T containing u. (see Figure 2.5 (a).)

Suppose that u belongs to directed cycles of T of all lengths k, $3 \le k \le p$, where $p < n$. We shall prove that there is a directed $(p + 1)$-cycle of T containing u.

Let C: $v_0, v_1, \ldots, v_{p-1}$ be a directed p-cycle containing u where $v_{p-1} = u$. Suppose that v is a vertex of T not belonging to C such that for some i and j, $0 \le i, j \le p - 1$, $i \ne j$, there exist arcs (v_i, v) and (v, v_j) of T. (See Figure 2.5 (b).) Then there must exist arcs (v_r, v) and (v, v_{r+1}) of $A(T)$, $i \le r \le j - 1$ (suffixes taken modulo p) and hence $(v_0, v_1, \ldots, v_r, v, v_{r+1}, \ldots, v_{p-1}, v_0)$ is a directed $(p + 1)$-cycle containing u. (See Figure 2.5 (b).)

If no such v exists, then for every vertex v of T not belonging to $V(C)$, either $(v_i, v) \in A(T)$ for every i, $0 \le i \le p - 1$, or $(v, v_i) \in A(T)$ for every i, $0 \le i \le$

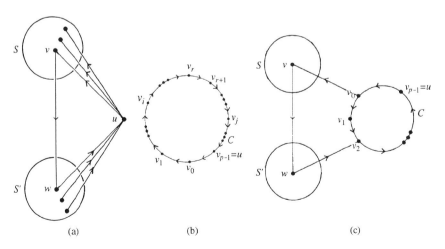

FIGURE 2.5. Graphs for proof of Theorem 2.2.2

$p - 1$. Let $S = \{v \in V(T)\backslash V(C) : (v_i, v) \in A(T)$ for each $i, 0 \leq i \leq p - 1\}$ and $S' = \{w \in V(T)\backslash V(C) : (w, v_i) \in A(T)$ for each $i, 0 \leq i \leq p - 1\}$. The diconnectedness of T implies that none of S, S', and $[S, S']$ is empty. Let (v, w) be an arc of $[S, S']$. Then $(v_0, v, w, v_2, \ldots, v_{p-1}, v_0)$ is a directed $(p + 1)$-cycle of T containing $v_{p-1} = u$. (See Figure 2.5 (c).) □

Remark 2.2.3 Theorem 2.2.2 shows, in particular, that every diconnected tournament is Hamiltonian, that is, it contains a directed Hamilton cycle.

Exercise 2.1 Show that every tournament T is diconnected or can be made into one by the reorientation of just one arc of T.

Exercise 2.2 Show that a tournament is diconnected if, and only if, it has a spanning directed cycle.

Exercise 2.3 Show that every tournament of order n has at most one vertex v with $d^+(v) = n - 1$.

Exercise 2.4 Show that for each positive integer $n \geq 3$, there exists a non-Hamiltonian tournament of order n.

Exercise 2.5 Show that if a tournament contains a directed cycle, then it contains a directed cycle of length 3.

Exercise 2.6 Show that every tournament T contains a vertex v such that every other vertex of T is reachable from v by a directed path of length at most 2.

2.3. k-Partite Tournaments

Definition 2.3.1 A *k-partite graph*, $k \geq 2$, is a graph G in which $V(G)$ is partitioned into k nonempty subsets $V_1, V_2, \ldots, V_k$, such that the induced subgraphs $G[V_1], G[V_2], \ldots, G[V_k]$ are all totally disconnected. It is said to be *complete* if, for $i \neq j$, each vertex of V_i is adjacent to every vertex of V_j, $1 \leq i, j \leq k$. A *k-partite tournament* is an oriented complete k-partite graph. (See Figure 2.6.)

The next three theorems are based on Goddard et al. [51]. We now give a characterization of a k-partite tournament containing a 3-cycle.

Theorem 2.3.2 *Let T be a k-partite tournament, $k \geq 3$. Then, T contains a directed 3-cycle if, and only if, there exists a directed cycle in T that contains vertices from at least three partite sets.*

Proof Suppose that T contains a directed 3-cycle C. Then the three vertices of C must belong to three distinct partite sets of T.

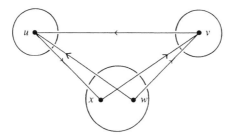

FIGURE 2.6. A 3-partite tournament

Conversely, suppose that T contains a directed cycle C that in turn contains vertices from at least three partite sets. Assume that C has the least possible length. Then there exist three consecutive vertices x, y, z on C that belong to distinct partite sets of T, say X, Y, Z. We claim that C is a directed 3-cycle. Assume that T has no directed 3-cycles. Then C is not a directed 3-cycle.

Because T is a tournament, either $(z, x) \in A(T)$ or $(x, z) \in A(T)$. If $(z, x) \in A(T)$, then (x, y, z, x) is a directed 3-cycle. But since C is a shortest directed cycle that contains vertices from three different partite sets, $C = (x, y, z, x)$, which is a contradiction to our assumption. Therefore $(x, z) \in A(T)$. Consider $C' = (C - y) + (x, z)$. C' is a directed cycle of length one less than that of C. So by assumption on C, C' contains vertices from only two partite sets, namely, X and Z. Let u be the vertex of C, immediately following z on C'. Then $u \in X$. If $(u, y) \in A(T)$, then $C_1'' = (y, z, u, y)$ is a directed 3-cycle containing vertices from three partite sets of T. Hence, by our assumption, $(u, y) \notin A(T)$, and as such $(y, u) \in A(T)$. (See Figure 2.7 (a) and (b).) Now consider $C_2'' = (C - z) + (y, u)$. C_2'' is a directed cycle that is shorter than C and contains at least one vertex other than x, y, and u. The successor of u in C_2'' belongs to Z, and thus C_2'' contains vertices from three partite sets of T. This is a contradiction to the choice of C. Thus (y, u) does not belong to $A(T)$ and consequently (x, z) does not belong to $A(T)$. Thus both (x, z) and (z, x) do not belong to $A(T)$, a contradiction. This proves the result. $\square$

Theorem 2.3.3 *Let T be a k-partite tournament, $k \geq 3$. Then every vertex u belonging to a directed cycle in T must belong to either a directed 3-cycle or a directed 4-cycle.*

Proof Let C be a shortest directed cycle in T that contains u. Suppose that C is not a directed 3-cycle. We shall prove that u is a vertex of a directed 4-cycle. Let u, x, y, and z be four consecutive vertices of C. If $(u, y) \in A(T)$, then $C' = (C - x) + (u, y)$ is a directed cycle in T containing u and having a shorter length than C. This contradicts the choice of C. Hence (u, y) does not belong to $A(T)$. Also if $(y, u) \in A(T)$, then (u, x, y, u) is a directed 3-cycle containing u. This contradicts our assumption on C. Hence $(y, u) \notin A(T)$. Consequently, y and u belong to the same partite sets of T. This means that u and z must belong to

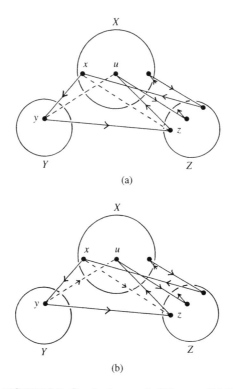

FIGURE 2.7. Graphs for proof of Theorem 2.3.2

distinct partite sets of T. If $(u, z) \in A(T)$, then $C'' = (C - \{x, y\}) + (u, z)$ is a directed cycle containing u and having length shorter than that of C. Hence (u, z) does not belong to $A(T)$. Therefore, $(z, u) \in A(T)$, and (u, x, y, z, u) is a directed 4-cycle containing u.   □

Remark 2.3.4   Theorem 2.2.2 states that every vertex of a diconnected tournament lies on a k-cycle for every $k, 3 \leq k \leq n$. However, this property is not true for a diconnected k-partite tournament. The tournament T of Figure 2.8 is a counterexample. T is a 3-partite tournament with $\{x, w\}$, $\{u\}$ and $\{v\}$ as partite sets. $(uwvxu)$ is a spanning directed cycle in T, and hence T is strongly connected but x is not a vertex of any directed 3-cycle.   □

Definition 2.3.5   The *score* of a vertex v in a tournament T is its outdegree. (This corresponds to the number of players who are beaten by player v). If $v_1, v_2, \ldots, v_n$ are the vertices of T and $s(v_i)$ is the score of v_i in T, then $(s(v_1), s(v_2), \ldots, s(v_n))$ is the *score vector* of T. An ordered triple (u, v, w) of vertices of T is a *transitive triple* of T if whenever $(u, v) \in A(T)$ and $(v, w) \in A(T)$, then $(u, w) \in A(T)$.

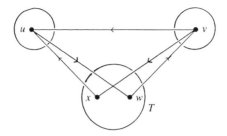

FIGURE 2.8. Diconnected 3-partite tournament T

Remarks 2.3.6

1. If v is any vertex of T and u, w are two outneighbors of v, then $\{u, v, w\}$ determines a unique transitive triple in T. Such a transitive triple is said to be defined by the vertex v. Clearly, any transitive triple of T is defined by some vertex of T. Further, the number of transitive triples defined by v is $\binom{s(v)}{2}$.

2. The number of directed 3-cycles in a tournament T of order n is obtained by subtracting the total number of transitive triples of vertices of T from the total number of triples of vertices of T. Thus the total number of directed 3-cycles in T is equal to $\binom{n}{3} - \sum_{v \in V(T)} \binom{s(v)}{2} = \frac{n(n-1)(n-2)}{6} - \frac{1}{2} \sum_{v \in V(T)} s(v)(s(v) - 1)$. Thus, the score vector of a tournament T determines the number of directed 3-cycles in T. But, in a general k-partite tournament, the score vector need not determine the number of directed 3-cycles. Consider the two 3-partite tournaments T_1 and T_2 of Figure 2.9. Both have the same score vector $(2, 2, 2, 2, 2, 2)$. But T_1 has eight directed 3-cycles whereas T_2 has only five directed 3-cycles.

Theorem 2.3.7 gives a formula for the number of directed 3-cycles in a k-partite tournament.

Theorem 2.3.7 *Let T be a k-partite tournament, $k \geq 3$, having partite sets $V_0, V_1, \ldots, V_{k-1}$. Then the number of directed 3-cycles in T is given by $\sum_{0 \leq i < j < l \leq k-1} |V_i||V_j||V_l| - \sum_{v \in V(T)} \sum_{i < j} O_i(v)O_j(v)$, where $O_i(v)$ denotes the number of outneighbors of v in V_i.*

Proof   Let S denote the set of triples of vertices of T such that the three vertices of the triple belong to three different partite sets, and let $N = |S|$. Then $N = \sum_{0 \leq i < j < l \leq k-1} |V_i||V_j||V_l|$.

Any orientation of a triangle gives a directed 3-cycle or a transitive triple. Hence, the number of directed 3-cycles in $T = N - N_1$, where N_1 is the number of transitive triples in T. Also a triple of vertices of T is transitive if, and only if, there exists a vertex of the triple having the other two as outneighbors. The number of such triples of T to which a vertex v can belong and for which the other two vertices are outneighbors of v is $\sum_{i < j} O_i(v)O_j(v)$. Hence, $N_1 = \sum_{v \in V(T)} \sum_{i < j} O_i(v)O_j(v)$. Thus, the number of directed 3-cycles in T is given by

$$N - N_1 = \sum_{0 \leq i < j < l \leq k-1} |V_i||V_j||V_l| - \sum_{v \in V(T)} \sum_{i < j} O_i(v)O_j(v). \qquad \square$$

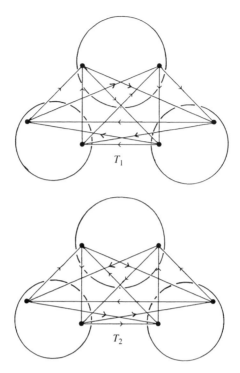

FIGURE 2.9. 3-partite tournaments with eight directed 3-cycles (T_1) and five directed 3-cycles (T_2)

Remark 2.3.8 For $k = 3$, the results of Theorem 2.3.7 simplify as follows:

(i) The number of transitive triples in a 3-partite tournament equals $\sum_{i=0}^{2} \sum_{v \in V_i} O_{i+1}(v) O_{i+2}(v)$, where the indices are taken modulo 3.

(ii) The number of directed 3-cycles in a 3-partite tournament T is given by $|V_0||V_1||V_2| - \sum_{i=0}^{2} \sum_{v \in V_i} O_{i+1}(v) O_{i+2}(v)$.

Remark 2.3.9 Consider the two 3-partite tournaments of Figure 2.10. T_1 has $|V_0||V_1||V_2|$ directed 3-cycles and has no transitive triples, whereas T_2 contains no directed 3-cycles and contains $|V_0||V_1||V_2|$ transitive triples.

Exercise 3.1 If $|V_i| = n_i$, $1 \leq i \leq k$, find the number of edges in the complete multipartite graph $G(V_1, V_2, \ldots, V_k)$. (See reference [19], p. 6.)

Exercise 3.2 If T is a strongly connected 3-partite tournament with partite sets V_0, V_1, V_2, then the maximum number of transitive triples in T is $|V_0||V_1||V_2| - 1$ unless $|V_0| = |V_1| = |V_2| = 2$, in which case T has at most $|V_0||V_1||V_2| - 2 = 6$ transitive triples.

Exercise 3.3 Construct a strongly connected 3-partite tournament containing exactly six transitive triples.

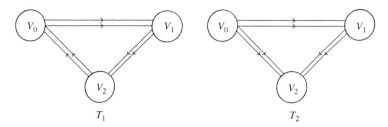

FIGURE 2.10. 3-partite tournaments T_1 (with directed 3-cycles and no transitive triples) and T_2 (with transitive triples and no directed 3-cycles). Double arrows indicate that all arcs joining corresponding partite sets have the same orientation.

Exercise 3.4 Give a definition of digraph isomorphism similar to that for graph isomorphism.

Exercise 3.5 Give an example of two nonisomorphic tournaments on five vertices. Justify your answer.

Exercise 3.6 If u and v are distinct vertices of a tournament T such that both $d(u, v)$ and $d(v, u)$ are defined (where $d(u, v)$ denotes a length of a shortest directed (u, v)-path in T), then show that $d(u, v) \neq d(v, u)$.

Notes

The earliest of the books on directed graphs is by Harary, Norman, and Cartwright [62]. *Topics on Tournaments* by Moon [91] deals exclusively with tournaments. Theorems 2.3.2 and 2.3.3 are based on reference [51].

III

Connectivity

3.0. Introduction

The connectivity of a graph is a "measure" of its connectedness. Some connected graphs are connected rather "loosely" in the sense that the deletion of a vertex or an edge from the graph destroys the connectedness of the graph. There are graphs in the other extreme as well, such as the complete graphs K_n, $n \geq 2$, which remain connected even after removal of all but one vertex.

Consider a communication network. Any such network can be represented by a graph in which the vertices correspond to communication centers and the edges represent communication channels. In the communication network of Figure 3.1(a), any disruption in the center v will result in a communication breakdown, whereas in the network of Figure 3.1(b), at least two centers have to be disrupted to cause a breakdown. It is needless to stress the importance of maintaining reliable communication networks at all times, especially during times of war. Hence the reliability of a communication network has a direct bearing on its connectivity.

In this chapter, we study the two graph parameters, namely, vertex connectivity and edge connectivity. We also introduce the parameter *cyclical edge connectivity*. We prove Menger's theorem and several of its variations. In addition, the theorem of Ford and Fulkerson on flows in networks is established.

3.1. Vertex Cuts and Edge Cuts

We now introduce the notions of vertex cuts, edge cuts, vertex connectivity, and edge connectivity.

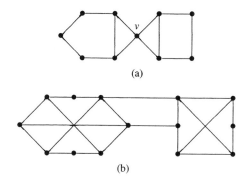

(a)

(b)

FIGURE 3.1. Two types of communication networks

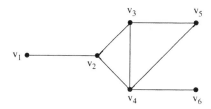

FIGURE 3.2. Graph illustrating vertex cuts and edge cuts

Definitions 3.1.1

1. A subset V' of the vertex set $V(G)$ of a connected graph G is a *vertex cut* of G, if $G-V'$ is disconnected; it is a *k-vertex cut* if $|V'| = k$. V' is then called a *separating set of vertices* of G. A vertex v of G is a *cut vertex* of G, if $\{v\}$ is a vertex cut of G.

2. Let G be a nontrivial connected graph with vertex set V and let S be a nonempty subset of V. For $\bar{S} = V \backslash S$, let $[S, \bar{S}]$ denote the set of all edges of G that have one end vertex in S and the other in $\bar{S}$. A set of edges of G of the form $[S, \bar{S}]$ is called an *edge cut* of G. An edge e is a *cut edge* of G, if $\{e\}$ is an edge cut of G. An edge cut of cardinality k is called a *k-edge cut* of G.

Example 3.1.2 For the graph of Figure 3.2, $\{v_2\}$, and $\{v_3, v_4\}$ are vertex cuts. The edge subsets $\{v_3v_5, v_4v_5\}$, $\{v_1v_2\}$, and $\{v_4v_6\}$ are all edge cuts. Of these, v_2 is a cut vertex, and v_1v_2 and v_4v_6 are cut edges. For the edge cut $\{v_3v_5, v_4v_5\}$, we may take $S = \{v_5\}$ so that $\bar{S} = \{v_1, v_2, v_3, v_4, v_6\}$.

Remarks 3.1.3

1. If uv is an edge of an edge cut E', then all the edges having u and v as their ends also belong to E'.

2. No loop can belong to an edge cut.

Exercise 1.1   If $\{x, y\}$ is a 2-edge cut of a graph G, show that every cycle of G that contains x must also contain y.

Remarks 3.1.4   If G is connected and E' is a set of edges whose deletion results in a disconnected graph, then E' contains an edge cut of G. It is clear that if e is a cut edge of a connected graph, then $G-e$ has exactly two components.

Remarks 3.1.5   Because the removal of a parallel edge of a connected graph does not result in a disconnected graph, such an edge cannot be a cut edge of the graph. A set of edges of a connected graph G whose deletion results in a disconnected graph is called a *separating set of edges*. In particular, any edge cut of a connected graph G is a separating set of edges of G.

We now characterize a cut vertex of G.

Theorem 3.1.6   *A vertex v of a connected graph G with at least three vertices is a cut vertex of G if, and only if, there exist vertices u and w of G, distinct from v, such that v is in every u–w path in G.*

Proof   If v is a cut vertex of G, then $G-v$ is disconnected and has at least two components, G_1 and G_2. Take $u \in V(G_1)$ and $w \in V(G_2)$. Then every u–w path in G must contain v, as otherwise u and w would belong to the same component of $G-v$. Conversely, suppose that the condition of the theorem holds. Then the deletion of v destroys every u–w path in G, and hence u and w lie in distinct components of $G-v$. Therefore, $G-v$ is disconnected and v is a cut vertex of G. $\square$

Theorems 3.1.7 and 3.1.8 characterize a cut edge of a graph.

Theorem 3.1.7   *An edge $e = xy$ of a graph G is a cut edge of a connected graph G if, and only if, e does not belong to any cycle of G.*

Proof   Let e be a cut edge of G, and let $[S, \bar{S}] = \{e\}$ be the partition of V defined by $G-e$ so that $x \in S$ and $y \in \bar{S}$. If e belongs to a cycle of G, then $[S, \bar{S}]$ must contain at least one more edge, contradicting that $\{e\} = [S, \bar{S}]$. Hence e cannot belong to a cycle.

Conversely, assume that e is not a cut edge of G. Then $G-e$ is connected, and hence there exists an x–y path P in $G-e$. Then $P \cup \{e\}$ is a cycle in G containing e. $\square$

Theorem 3.1.8   *An edge $e = xy$ is a cut edge of a connected graph G if, and only if, there exist vertices u and v such that e belongs to every u–v path in G.*

Proof   Let $e = xy$ be a cut edge of G. Then $G-e$ has two components, G_1 and G_2. Let $u \in V(G_1)$ and $v \in V(G_2)$. Then clearly, every u–v path in G contains e.

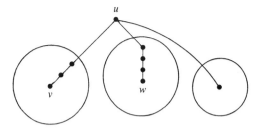

FIGURE 3.3. Graph for proof of Theorem 3.1.10

Conversely, suppose that there exist vertices u and v satisfying the condition of the theorem. Then, there exists no u–v path in G–e so that G–e is disconnected. Hence e is a cut edge of G. □

Remark 3.1.9 There exist graphs in which every edge is a cut edge. It follows from Theorem 3.1.7 that if G is a simple connected graph with at least one edge and without cycles, then every edge of G is a cut edge of G. A similar result is not true for cut vertices. Our next result shows that not every vertex of a connected graph (with at least two vertices) is a cut vertex of G.

Theorem 3.1.10 *A connected graph G with at least two vertices contains at least two vertices that are not cut vertices.*

Proof First suppose that $n(G) \geq 3$. Let u and v be vertices of G such that $d(u, v)$ is maximum. Then neither u nor v is a cut vertex of G, because if u were a cut vertex of G, G–u would be disconnected, having at least two components. The vertex v belongs to one of these components. Let w be any vertex belonging to a component of G–u not containing v. Then every v–w path in G must contain u (see Figure 3.3). Consequently $d(v, w) > d(v, u)$, contradicting the choice of u and v. Hence, u is not a cut vertex of G. Similarly, v is not a cut vertex of G.

If $n(G) = 2$, then K_2 is a spanning subgraph of G, and so no vertex of G is a cut vertex of G. This completes the proof of the theorem. □

Proposition 3.1.11 *A simple cubic (i.e., 3-regular) connected graph G has a cut vertex if, and only if, it has a cut edge.*

Proof Let G have a cut vertex v_0. Let v_1, v_2, v_3 be the vertices of G that are adjacent to v_0 in G. Consider G–v_0, which has either two or three components. If G–v_0 has three components, no two of v_1, v_2, and v_3 can belong to the same component of G–v_0. In this case, each of v_0v_1, v_0v_2, and v_0v_3 is a cut edge of G. (See Figure 3.4 (a).) In the case when G–v_0 has only two components, one of the vertices, say v_1, belongs to one component of G–v_0, and v_2 and v_3 belong to the other component. In this case v_0v_1 is a cut edge. (See Figure 3.4 (b).)

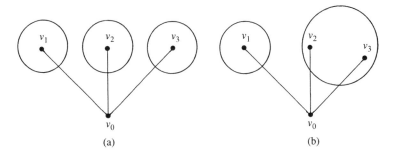

FIGURE 3.4. Graph for proof of Proposition 3.1.11

Conversely, suppose that $e = uv$ is a cut edge of G. Then the deletion of u results in the deletion of the edge uv. Since G is cubic, $G-u$ is disconnected. Accordingly, u is a cut vertex of G. $\square$

Exercise 1.2   Find the vertex cuts and edge cuts of the graph of Figure 3.2.

Exercise 1.3   Prove or disprove: Let G be a simple connected graph with $n(G) \geq 3$. Then G has a cut edge if, and only if, G has a cut vertex.

Exercise 1.4   Show that in a graph, the number of edges common to a cycle and an edge cut is even.

3.2. Connectivity and Edge-Connectivity

We now introduce two parameters of a graph, which, in a way, measure the connectedness of the graph.

Definition 3.2.1   For a nontrivial connected graph G having a pair of nonadjacent vertices, the minimum k for which there exists a k-vertex cut is called the *vertex connectivity* or simply the *connectivity* of G; it is denoted by $\kappa(G)$ or simply κ (kappa) when G is understood. If G is trivial or disconnected, $\kappa(G)$ is taken to be zero, whereas if G contains K_n as a spanning subgraph, $\kappa(G)$ is taken to be $n - 1$.

A set of vertices or edges of a connected graph G is said to *disconnect* the graph if its deletion results in a disconnected graph.

When a connected graph G (on n vertices) does not contain K_n as a spanning subgraph, κ is the connectivity of G if there exists a set of κ vertices of G whose deletion results in a disconnected subgraph of G while no set of $\kappa - 1$ (or fewer) vertices has this property.

Exercise 2.1   Prove that a simple graph G with n vertices, $n \geq 2$, is complete if, and only if, $\kappa(G) = n - 1$.

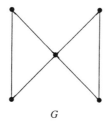

G

FIGURE 3.5. A1-connected graph G

Definition 3.2.2 The *edge connectivity* of a connected graph G is the smallest k for which there exists a k-edge cut (i.e., an edge cut containing k edges). The edge connectivity of a trivial or disconnected graph is taken to be 0. The edge connectivity of G is denoted by $\lambda(G)$. If λ is the edge connectivity of a connected graph G, there exists a set of λ edges whose deletion results in a disconnected graph, and no subset of edges of G of size less than λ has this property.

Exercise 2.2 Prove that the deletion of edges of a minimum edge cut of a connected graph G results in a disconnected graph with exactly two components. (Note that a similar result is not true for a minimum vertex cut.)

Definition 3.2.3 A graph G is *r-connected* if $\kappa(G) \geq r$. G is *r-edge connected* if $\lambda(G) \geq r$.

An r-connected (respectively, r-edge connected) graph is also ℓ-connected (respectively, ℓ-edge connected) for each ℓ, $0 \leq \ell \leq r - 1$.
    For the graph G of Figure 3.5, $\kappa(G) = 1$ and $\lambda(G) = 2$.
    We now derive inequalities connecting $\kappa(G)$, $\lambda(G)$, and $\delta(G)$.

Theorem 3.2.4 *For any loopless connected graph G, $\kappa(G) \leq \lambda(G) \leq \delta(G)$.*

Proof $\kappa = 0$ if, and only if, $\lambda = 0$. Also, $\delta = 0$ implies $\kappa = 0$ and $\lambda = 0$. Hence we may assume that κ, λ, and δ are all at least 1. Let E be an edge cut of G with λ edges. Let u and v be the end vertices of an edge of E . For each edge of E that does not have both u and v as end vertices, remove an end vertex that is different from u and v. If there are t such edges, at most t vertices have been removed. If the resulting graph is disconnected, then $\kappa \leq t < \lambda$. Otherwise, there will remain a subset of edges of E having u and v as end vertices, the removal of which would disconnect the graph. Hence, in addition to the already removed vertices, the removal of one of u and v will result in either a disconnected graph or a trivial graph. In the process, a set of at most $t + 1$ vertices has been removed and $\kappa \leq t + 1 \leq \lambda$.
    Finally, it is clear that $\lambda \leq \delta$. In fact, if v is a vertex of G with $d_G(v) = \delta$, then the set of δ edges of G incident at $v = ([\{v\}, V \backslash \{v\}])$ forms an edge cut of G. Thus $\lambda \leq \delta$. □

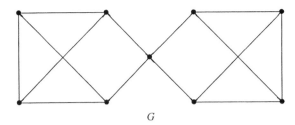

FIGURE 3.6. Graph G with $\kappa = 1, \lambda = 2$, and $\delta = 3$

It is possible that the inequalities in Theorem 3.2.4 can be strict. See the graph G of Figure 3.6, for which $\kappa = 1, \lambda = 2$, and $\delta = 3$.

Exercise 2.3 Prove or disprove: If H is a subgraph of G, then

(a) $\kappa(H) \le \kappa(G)$ and
(b) $\lambda(H) \le \lambda(G)$.

Exercise 2.4 Determine $\lambda(K_n)$.

Exercise 2.5 Determine the connectivity and edge-connectivity of the Petersen graph P. (See graph P of Figure 1.7. Note that P is a cubic graph.)

Theorem 3.2.5 gives a class of graphs for which $\kappa = \lambda$.

Theorem 3.2.5 *The connectivity and edge connectivity of a simple cubic graph G are equal.*

Proof We need only consider the case of a connected cubic graph. Again, since $\kappa \le \lambda \le \delta = 3$, we have only to consider the cases when $\kappa = 1, 2$, or 3. Now, Proposition 3.1.11 implies that for a simple cubic graph G, $\kappa = 1$ if, and only if, $\lambda = 1$.

If $\kappa = 3$, then by Theorem 3.2.4, $3 = \kappa \le \lambda \le \delta = 3$, and hence $\lambda = 3$.

We shall now prove that $\kappa = 2$ implies that $\lambda = 2$.

Suppose $\kappa = 2$ and $\{u, v\}$ is a 2-vertex cut of G. The deletion of $\{u, v\}$ results in a disconnected subgraph G' of G. Since each of u and v must be joined to each component of G', and since G is cubic, G' can have at most three components. If G' has three components, G_1, G_2, and G_3, and if e_i and $f_i, i = 1, 2, 3$, join respectively u and v with G_i, then each pair $\{e_i, f_i\}$ is an edge cut of G. (See Figure 3.7 (a).)

If G' has only two components, G_1 and G_2, then each of u and v is joined to one of G_1 and G_2 by a single edge, say, e and f, respectively, so that $\{e, f\}$ is an edge cut of G. (See Figures 3.7 (b), (c), and (d).)

Hence in either case there exists an edge cut consisting of two edges. As such, $\lambda \le 2$. But by Theorem 3.2.4, $\lambda \ge \kappa = 2$. Hence $\lambda = 2$. Finally, the above arguments show that if $\lambda = 3$, then $\kappa = 3$, and if $\lambda = 2$, then $\kappa = 2$. □

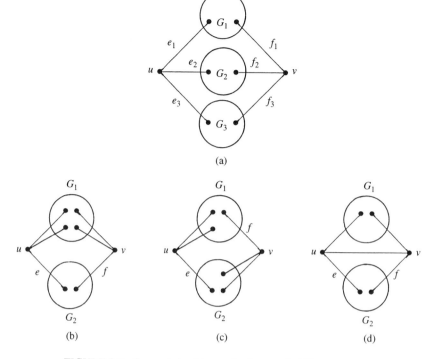

FIGURE 3.7. Connected cubic graphs for proof of Theorem 3.2.5

Exercise 2.6   Give examples of cubic graphs G_1, G_2, and G_3 with $\kappa(G_1) = 1, \kappa(G_2) = 2$, and $\kappa(G_3) = 3$.

Definition 3.2.6   A family of two or more paths in a graph G is said to be *internally disjoint* if no vertex of G is an internal vertex of more than one path in the family.

We now state and prove *Whitney's theorem* on 2-connected graphs.

Theorem 3.2.7 (Whitney)   *A graph G with at least three vertices is 2-connected if, and only if, any two vertices of G are connected by at least two internally disjoint paths.*

Proof   Let G be 2-connected. Then G contains no cut vertex. Let u and v be two distinct vertices of G. We now use induction on $d(u, v)$ to prove that u and v are joined by two internally disjoint paths.

If $d(u, v) = 1$, let $e = uv$. As G is 2-connected and $n(G) \geq 3$, e cannot be a cut edge of G, since if e were a cut edge, at least one of u and v must be a cut vertex. By Theorem 3.1.7, e belongs to a cycle C in G. Then $C-e$ is a u–v path in G, internally disjoint from the path uv.

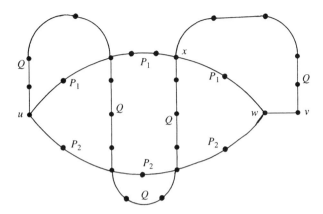

FIGURE 3.8. Graph for proof of Theorem 3.2.7

Now assume that any two vertices x and y of G, such that $d(x, y) = k-1, k \geq 2$, are joined by two internally disjoint x–y paths in G. Let $d(u, v) = k$. Let P be a u–v path of length k and w be the vertex of G just preceding v on P. Then $d(u, w) = k - 1$. By induction hypothesis, there are two internally disjoint u–w paths, say P_1 and P_2, in G. As G has no cut vertex, $G-w$ is connected and hence there exists a u–v path Q in $G-w$. Q is clearly a u–v path in G not containing w. Let x be the vertex of Q such that the x–v section of Q contains only the vertex x in common with $P_1 \cup P_2$. (See Figure 3.8.)

We may suppose, without loss of generality, that x belongs to P_1. Then the union of the u–x section of P_1 and x–v section of Q and $P_2 \cup (wv)$ are two internally disjoint u–v paths in G. This gives the proof in one direction.

In the other direction, assume that any two distinct vertices of G are connected by at least two internally disjoint paths. Then G is connected. Further, G cannot contain a cut vertex since, if v were a cut vertex of G, there must exist vertices u and w such that every u–w path contains v (compare Theorem 3.1.6), contradicting the hypothesis. Hence G is 2-connected. □

Theorem 3.2.8 *A graph G with at least three vertices is 2-connected if and only if, any two vertices of G lie on a common cycle.*

Proof Let u and v be any two vertices of a 2-connected graph G. By Theorem 3.2.7, there exist two internally disjoint paths in G joining u and v. The union of these two paths is a cycle containing u and v.

Conversely, if any two vertices u and v lie on a cycle C, then C is the union of two internally disjoint u–v paths. Again, by Theorem 3.2.7, G is 2-connected.

□

Remark 3.2.9 If G is 2-connected, and if u and v are distinct vertices of G, and P, a u–v path in G, it is not in general true that there exists another u–v path Q in G such that P and Q are internally disjoint. For example, in the 2-connected graph of Figure 3.9, if P is the u–w' path $uwvv'u'w'$, then there exists no u–w' path

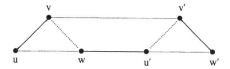

FIGURE 3.9. Graph for Remark 3.2.9

Q in G that is internally disjoint from P. However, there do exist two internally disjoint u–w' paths in G.

Exercise 2.7

(a) Show that a graph G with at least three vertices is 2-connected if, and only if, any two edges of G lie on a common cycle.

(b) Show that a graph G with at least three vertices is 2-connected if, and only if, any vertex and any edge of G lie on a common cycle of G.

Exercise 2.8   Prove that a graph is 2-connected if, and only if, for every pair of disjoint connected subgraphs G_1 and G_2, there exist two internally disjoint paths P_1 and P_2 of G between G_1 and G_2.

Exercise 2.9   *Edge form of Whitney's Theorem:* Prove that a graph G with $n \geq 3$ is 2-edge connected if, and only if, any two distinct vertices of G are connected by at least two internally edge-disjoint paths in G. (Hint: Imitate the proof of Theorem 3.2.7, or pass on to $L(G)$.)

Exercise 2.10

(a) Disprove by a counterexample: If $\kappa(G) = k$, then $\kappa(L(G)) = k$.

(b) Prove: $\lambda(G) \leq \kappa(L(G))$. Give an example of a graph G for which $\lambda(G) < \kappa(L(G))$.

Theorem 3.2.10   *In a 2-connected graph G, any two longest cycles have at least two vertices in common.*

Proof   Let $C_1 = u_1 u_2 \cdots u_k u_1$ and $C_2 = v_1 v_2 \cdots v_k v_1$ be two longest cycles in G. If C_1 and C_2 are disjoint, there exist (since G is 2-connected) two disjoint paths, say P_1 joining u_i and v_j and P_2 joining u_ℓ and v_m, connecting C_1 and C_2 such that $u_i \neq u_\ell$ and $v_j \neq v_m$ (see Exercise 2.8). u_i and u_ℓ divide C_1 into two subpaths. Let L_1 be the longer of these subpaths. (If both the subpaths are of equal lenght, we take either one of them to be L_1.) Let L_2 be defined in a similar manner in C_2. Then $L_1 \cup P_1 \cup L_2 \cup P_2$ is a cycle of length greater than that of C_1 (or C_2). Hence C_1 and C_2 cannot be disjoint. (See Figure 3.10.)

Suppose that C_1 and C_2 have exactly one vertex, say $u_1 = v_1$, in common. Since G is 2-connected, u_1 is not a cut vertex of G, and so there exists a path P with one end vertex u_i in C_1–u_1 and the other end vertex v_j in C_2–v_1, which is internally disjoint from $C_1 \cup C_2$. Let P_1 denote the longer of the two u_1–u_i sections of C_1, and Q_1 denote the longer of the two v_1–v_j sections of C_2. Then $P_1 \cup P \cup Q_1$ is a

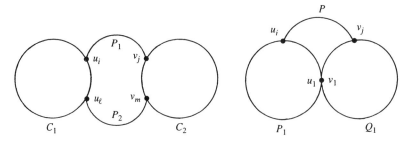

FIGURE 3.10. Graphs for proof of Theorem 3.2.10

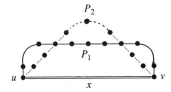

FIGURE 3.11. Graph for proof of Theorem 3.2.11

cycle longer than C_1 (or C_2). But this is impossible. Thus C_1 and C_2 must have at least two vertices in common. $\square$

Theorem 3.2.11 gives a simple characterization of 3-edge connected graphs.

Theorem 3.2.11 *A connected simple graph G is 3-edge connected if, and only if, every edge of G is the (exact) intersection of the edge sets of two cycles of G.*

Proof Let G be 3-edge connected and let $x = uv$ be an edge of G. Since $G - x$ is 2-edge connected, there exist two edge-disjoint $u–v$ paths P_1 and P_2 in $G - x$ (see Exercise 2.9). Now, $P_1 \cup \{x\}$ and $P_2 \cup \{x\}$ are two cycles of G, the intersection of whose edge sets is precisely $\{x\}$. (See Figure 3.11).

Conversely, suppose that for each edge $x = uv$ there exist two cycles C and C' such that $\{x\} = E(C) \cap E(C')$. G cannot have a cut edge, since by hypothesis each edge belongs to two cycles and no cut edge can belong to a cycle; nor can G contain an edge cut consisting of two edges x and y, by Exercise 1.1. Hence, $\lambda(G) \geq 3$, and G is 3-edge connected. $\square$

3.3. Blocks

In this section, we specialize in connected graphs without cut vertices.

Definition 3.3.1 A graph G is *nonseparable* if it is nontrivial, connected, and has no cut vertices. A *block of a graph* is a maximal nonseparable subgraph of G. If G has no cut vertex, G itself is a *block*.

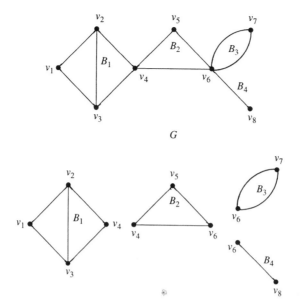

FIGURE 3.12. A graph G and its blocks

In Figure 3.12, a graph G and its blocks B_1, B_2, B_3 and B_4 are indicated. B_1, B_3, and B_4 are the *end blocks* of G (i.e., blocks having exactly one cut vertex of G). The following facts are worthy of observation.

Remarks 3.3.2 Let G be a connected graph with $n \geq 3$.

1. Each block of G with at least three vertices is a 2-connected subgraph of G.

2. Each edge of G belongs to one of its blocks. Hence, G is the union of its blocks.

3. Any two blocks of G have at most one vertex in common. (Such a common vertex is a cut vertex of G).

4. A vertex of G that is not a cut vertex belongs to exactly one of its blocks.

5. A vertex of G is a cut vertex of G if, and only if, it belongs to at least two blocks of G.

Whitney's Theorem (Theorem 3.2.7) implies that a graph with at least three vertices is a block if, and only if, any two vertices of the graph are connected by at least two internally disjoint paths. Again, by Theorem 3.2.8, we see that any two vertices of a block with at least three vertices belong to a common cycle. Thus, a block with at least three vertices contains a cycle.

Theorem 3.3.3 (Ear decomposition of a block) *If C is any cycle of a simple block G with at least three vertices, then there exists a sequence of nonseparable subgraphs $C = B_0, B_1, \ldots, B_r = G$ such that B_{i+1} is an edge-disjoint union*

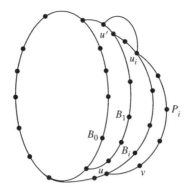

FIGURE 3.13. Graph for proof of Theorem 3.3.3

of B_i and a path P_i, where the only vertices common to B_i and P_i are the end vertices of P_i, $0 \leq i \leq r - 1$.

Proof Assume that we have already determined B_i (see Figure 3.13). If $B_i \neq G$, there exists (as G is connected) an ẽdge $e = uv$ not belonging to B_i, but with u in B_i. If v also belongs to B_i, take $P_i = uv$ and $B_{i+1} = B_i \cup P_i$. Otherwise $e = uv$ is an edge of G having only one of its ends, namely u, in B_i. Let u' be any other vertex of B_i. Then, since G is 2-connected, e and u' belong to a common cycle C_i (see Exercise 2.7). Let u_i be the first vertex of B_i in the u–u' section C' of C_i containing v, and let P_i be the u–u_i section of C'. Define $B_{i+1} = B_i \cup P_i$. Then B_{i+1} is nonseparable, and the proof follows by induction on i. □

3.4. Cyclical Edge-Connectivity of a Graph

In this section we introduce the parameter, cyclical edge connectivity of graphs. Unlike connectivity and edge connectivity, cyclical edge connectivity is not defined for all graphs.

Definition 3.4.1 Let G be a connected simple graph containing at least two disjoint cycles. Then the *cyclical edge connectivity* of G is defined to be the minimum number of edges of G whose deletion results in a graph having two components, each containing a cycle. It is denoted by $\lambda_c(G)$.

It is clear that $\lambda \leq \lambda_c$. The graphs G and H of Figure 3.14 show that both $\lambda = \lambda_c$ and $\lambda < \lambda_c$ can happen.

If a connected graph G is a union of edge-disjoint cycles, then $\lambda_c(G)$ is even. This is because any edge cut has an even number of edges in common with a cycle (see Exercise 1.4), and in the present case every edge belongs to a cycle.

Exercise 4.1 Show that the cyclical edge connectivity of the Petersen graph P is 5. (Hint: P has 10 vertices, 15 edges, and *girth* (= length of a smallest cycle) 5.)

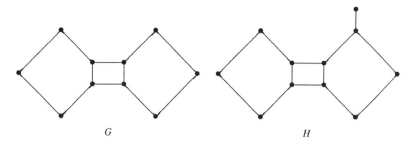

FIGURE 3.14. Graphs $G(\lambda(G) = \lambda_c(G) = 2)$ and $H(\lambda(H) = 1, \lambda_c(H) = 2)$

3.5. Menger's Theorem

In this section we prove different versions of the celebrated Menger's Theorem, which generalizes Whitney's Theorem (Theorem 3.2.7). Menger's Theorem [88] relates the connectivity of a graph G to the number of internally disjoint paths between pairs of vertices of G. The proofs given here make use of network analysis. We begin with the definition of a network.

Definition 3.5.1 A *network* N is a digraph D with two distinguished vertices s and t, $(s \neq t)$, and a nonnegative integer-valued function c defined on its arc set A. s is called the *source* and t is called the *sink* of N. The source corresponds to the supply center and the sink corresponds to a market. Vertices of N, other than s and t, are called the *intermediate vertices* of N. The digraph D is called the underlying digraph of N. The function c is called the *capacity function* of N, and $c(a)$, for an arc a, denotes the *capacity* of a.

Example 3.5.2 A network N is diagrammatically represented by the underlying digraph D, labeling each arc with its capacity. Figure 3.15 is a network with source s and sink t, and three intermediate vertices. The numbers inside the brackets denote the capacities of the respective arcs.

For a real-valued function f defined on A, and $K \subseteq A$, $\sum_{a \in K} f(a)$ will be denoted by $f(K)$. If K is a set of arcs of D of the form $[S, \bar{S}]$, that is, the set of arcs with heads in $\bar{S}$ and tails in S, where $S \subseteq V(D)$, $\bar{S} = V(D) \backslash S$, then $f^+(S)$ and $f^-(S)$ denote $f([S, \bar{S}])$ and $f([\bar{S}, S])$, respectively. If $S = \{v\}$, then $f^+(S)$ and $f^-(S)$ are denoted by $f^+(v)$ and $f^-(v)$, respectively.

Definition 3.5.3 A *flow* in a network N is an integer-valued function f defined on $A = A(N)$ such that $0 \leq f(a) \leq c(a)$ for all $a \in A$ and $f^+(v) = f^-(v)$ for all the intermediate vertices v of N.

Remarks 3.5.4

1. $f^+(v)$ is the flow out of v, and $f^-(v)$ is the flow into v. *The condition $f^+(v) = f^-(v)$ for each intermediate vertex v then signifies that there is conservation of flow at every such vertex.*

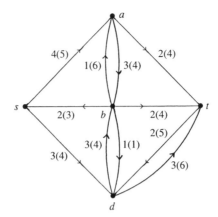

FIGURE 3.15. Network with source (s), sink (t), and three intermediate vertices

2. If $a = (u, v)$, we denote $f(a)$ by f_{uv}. Every network N has at least one flow, since the function f defined by $f(a) = 0$ for all $a \in A$ is a flow. It is called the *zero flow* in N.

3. A less trivial example of a flow in the network of Figure 3.15 is given by f, where $f_{sa} = 4$, $f_{sd} = 3$, $f_{bs} = 2$, $f_{at} = 2$, $f_{ab} = 3$, $f_{ba} = 1$, $f_{bd} = 1$, $f_{db} = 3$, $f_{bt} = 2$, $f_{dt} = 3$, $f_{td} = 2$.

4. If S is a subset of vertices in a network N and f is a flow in N, $f^+(S) - f^-(S)$, is called the *resultant flow out of S* and $f^-(S) - f^+(S)$, the *resultant flow into S*, relative to f.

5. The flow along any arc (u, v) is both the outflow at u along (u, v) and the inflow at v along (u, v). Hence $\sum_{v \in V(N)} f^+(v) - \sum_{v \in V(N)} f^-(v) = 0$. This gives that

$$[f^+(s) - f^-(s)] + \sum_{\substack{v \in V(N) \\ v \neq s, t}} (f^+(v) - f^-(v)) + [f^+(t) - f^-(t)] = 0.$$

But $f^+(v) = f^-(v)$ for each $v \in V(N)$, $v \neq s, t$. Hence, $f^+(s) - f^-(s) = f^-(t) - f^+(t)$. Thus, relative to any flow f, the resultant flow out of s is equal to the resultant flow into t. For a similar reason, if S is any subset of $V(N)$ containing s but not t.

$$\sum_{v \in S} f^+(v) - \sum_{v \in S} f^-(v) = f^+(s) - f^-(s). \qquad (*)$$

This common quantity is called the *value* of f and is denoted by val f. Thus, val $f = f^+(s) - f^-(s) = f^-(t) - f^+(t)$. The value of the flow f of the network of Figure 3.15 is 5.

Definition 3.5.5

1. A flow f in N is a *maximum flow* if there is no flow f' in N such that val $f' > $ val f.

2. A *cut* in N is a set of arcs of the form $[S, \bar{S}]$ where $s \in S$ and $t \in \bar{S}$. Such a cut is said to *separate* s and t. For example, $\{(a, t), (b, t), (d, t)\}$ is a cut in the network of Figure 3.15, where $S = \{s, a, b, d\}$.

3. The *capacity* of a cut K is the sum of the capacities of its arcs. We denote the capacity of K by cap K. Thus, cap $K = \sum_{a \in K} c(a)$.

Theorem 3.5.6 gives the relation between the value of a flow and the capacity of a cut in a network.

Theorem 3.5.6 *In any network N, the value of any flow f is less than or equal to the capacity of any cut K.*

Proof Let $[S, \bar{S}]$ be any cut. We have, by (*),

$$\text{val } f = f^+(s) - f^-(s)$$

$$= \sum_{v \in S} f^+(v) - \sum_{v \in S} f^-(v)$$

$$= \sum_{\substack{v \in S \\ u \in S \\ (v,u) \in A(D)}} f_{vu} + \sum_{\substack{v \in S \\ u \in \bar{S} \\ (v,u) \in A(D)}} f_{vu} - \sum_{\substack{u \in S \\ v \in S \\ (u,v) \in A(D)}} f_{uv} - \sum_{\substack{u \in \bar{S} \\ v \in S \\ (u,v) \in A(D)}} f_{uv}.$$

But

$$\sum_{\substack{v \in S \\ u \in S \\ (v,u) \in A(D)}} f_{vu} - \sum_{\substack{u \in S \\ v \in S \\ (u,v) \in A(D)}} f_{uv} = 0.$$

Hence

$$\text{val } f = \sum_{\substack{v \in S \\ u \in \bar{S} \\ (v,u) \in A(D)}} f_{vu} - \sum_{\substack{u \in \bar{S} \\ v \in S \\ (u,v) \in A(D)}} f_{uv}.$$

Since

$$\sum_{\substack{u \in \bar{S} \\ v \in S \\ (u,v) \in A(D)}} f_{uv} \geq 0$$

(recall that f is a nonnegative integer-valued function), we get

$$\text{val } f \leq \sum_{\substack{v \in S \\ u \in \bar{S} \\ (v,u) \in A(D)}} f_{vu} \leq \sum_{\substack{v \in S \\ u \in \bar{S} \\ (v,u) \in A(D)}} c_{(v,u)} = c([S, \bar{S}]). \qquad \square$$

Note 3.5.7 Note that we have shown that val f is the flow out of S minus the flow into S for any $S \subset V$ with $s \in S$ and $t \in \bar{S}$.

By Theorem 3.5.6, the value of any flow f does not exceed the capacity of any cut K. In particular, if f^* is a maximum flow in N and K^* is a minimum cut, that is, a cut with minimum capacity, then val $f^* \leq$ cap K^*.

Lemma 3.5.8 *Let f be a flow and K a cut in a network N such that val $f =$ cap K. Then f is a maximum flow and K is a minimum cut.*

Proof Let f^* be a maximum flow, and K^* be a minimum cut in N. Then we have, by Theorem 3.5.6, val $f \leq$ val $f^* \leq$ cap $K^* \leq$ cap K. But by hypothesis, val $f =$ cap K. Hence, val $f =$ val $f^* =$ cap $K^* =$ cap K. Thus f is a maximum flow and K is a minimum cut. □

Theorem 3.5.9 is the celebrated *max-flow min-cut theorem* due to Ford and Fulkerson [42], which establishes the equality of the value of a maximum flow and the minimum capacity of a cut separating s and t.

Theorem 3.5.9 (Ford and Fulkerson) *In a given network N with source s and sink t, the maximum value of a flow from s to t is equal to the minimum value of the capacities of all the cuts in N.*

Proof In view of Lemma 3.5.8, we need only prove that there exists a flow in N whose value is equal to $c([S, \bar{S}])$ for some cut $[S, \bar{S}]$ separating s and t. Let f be a maximum flow in N with val $f = w_0$. Define $S \subset N$ recursively as follows:

(a) $s \in S$, and
(b) if a vertex $u \in S$ and either $f_{uv} < c(u, v)$ or $f_{vu} > 0$, then include v in S.

Any vertex not belonging to S belongs to $\bar{S}$. We claim that t cannot belong to S, for, if we suppose that $t \in S$, then there exists a path P from s to t, say P: $s v_1 v_2 \cdots v_j v_{j+1} \cdots v_k t$, with its vertices in S such that for any arc of P, either $f_{v_j v_{j+1}} < c(v_j, v_{j+1})$ or $f_{v_{j+1} v_j} > 0$. Call an arc joining v_j and v_{j+1} of P a *forward arc*, if it is directed from v_j to v_{j+1}; otherwise, it is a *backward arc*.

Let δ_1 be the minimum of all differences $(c(v_j, v_{j+1}) - f_{v_j v_{j+1}})$ for forward arcs, and let δ_2 be the minimum of all flows in backward arcs of P. Both δ_1 and δ_2 are positive, by the definition of S. Let $\delta = \min\{\delta_1, \delta_2\}$. Increase the flow in each forward arc of P by δ and also decrease the flow in each backward arc of P by δ. Keep the flows along the other arcs of N unaltered. Then there results a new flow whose value is $w_0 + \delta > w_0$, leading to a contradiction. (This is because among all arcs incident at s, only in the initial arc of P is the flow value increased by δ if it is a forward arc or decreased by δ if it is a backward arc; see (5) of Remarks 3.5.4.) This contradiction shows that $t \notin S$, and therefore $t \in \bar{S}$. In other words, $[S, \bar{S}]$ is a cut separating s and t. If $v \in S$ and $u \in \bar{S}$, we have, by the definition of S, $f_{vu} = c(v, u)$ if (v, u) is an arc of N, and $f_{uv} = 0$ if (u, v) is an arc of N.

Hence, as in the proof of Theorem 3.5.6,

$$w_0 = \sum_{\substack{u \in S \\ v \in \bar{S}}} f_{uv} - \sum_{\substack{v \in \bar{S} \\ u \in S}} f_{vu} = \sum_{\substack{u \in S \\ v \in \bar{S}}} f_{uv} - 0 = \sum_{\substack{u \in S \\ v \in \bar{S}}} c(u, v) = c([S, \bar{S}]). \qquad \square$$

We now use the max-flow min-cut theorem to prove a number of results due to Menger. We shall first prove a result for a network in which each arc has unit capacity.

Theorem 3.5.10 *Let N be a network with source s and sink t. Let each arc of N have unit capacity. Then,*

(a) *the value of a maximum flow in N is equal to the maximum number k of arc-disjoint directed (s, t)-paths in N, and*

(b) *the capacity of a minimum cut in N is equal to minimum number ℓ of arcs whose deletion destroys all (s, t)-paths in N.*

Proof Let f^* be a maximum flow in N, and let D^* denote the digraph obtained from D by deleting all arcs whose flow is zero in f^*. Now, note that $0 < f^*(a) \leq c(a) = 1$, for all $a \in A(D^*)$, and therefore, $f^*(a) = 1$ for all $a \in A(D^*)$. Hence,

(i) $d_{D^*}^+(s) - d_{D^*}^-(s) = \text{val } f^* = d_{D^*}^-(t) - d_{D^*}^+(t)$ and

(ii) $d_{D^*}^+(v) = d_{D^*}^-(v)$ for $v \in V(N) \setminus \{s, t\}$.

(i) and (ii) imply that there are val f^* arc-disjoint directed (s, t)-paths in D^*, and hence also in D. Thus val $f^* \leq k$. Now let $P_1, P_2, \ldots, P_k$ be any system of k arc-disjoint directed (s, t)-paths in N. Define a function f on $A(N)$ by

$$f(a) = \begin{cases} 1 & \text{if } a \text{ is an arc of } \bigcup_{i=1}^{k} P_i, \\ 0 & \text{otherwise.} \end{cases}$$

Then f is a flow in N with value k. Since f^* is a maximum flow, we have val $f^* \geq k$. Consequently, val $f^* = k$, proving (a).

Let $K^* = [S, \bar{S}]$ be a minimum cut in N so that $|K^*| \geq \ell$, by the definition of ℓ. Then, cap $K^* = |K^*| \geq \ell$.

Now let Z be a set of ℓ arcs whose deletion destroys all directed (s, t)-paths, and let T denote the set of all vertices joined to s by a directed path in $N-Z$. Then since $s \in T$, and $t \in \bar{T}$, $K = [T, \bar{T}]$ is a cut in N. By definition of T, $N-Z$ can contain no arc of $[T, \bar{T}]$, and hence $K \subseteq Z$. Since K^* is a minimum cut, we conclude that cap $K^* \leq$ cap $K = |K| \leq |Z| = \ell$. Thus, cap $K^* = \ell$. $\square$

We now state and prove the *edge version of Menger's theorem for directed graphs*.

Theorem 3.5.11 *Let x and y be two vertices of a digraph D. Then the maximum number of arc-disjoint directed (x, y)-paths in D is equal to the minimum number of arcs whose deletion destroys all directed (x, y)-paths in D.*

Proof   Apply Theorem 3.5.9 to the two results of Theorem 3.5.10.   □

Theorem 3.5.12 is the *edge version of Menger's theorem for undirected graphs.*

Theorem 3.5.12   *Let x and y be two vertices of a graph G. Then the maximum number of edge-disjoint (x, y)-paths in G is equal to the minimum number of edges of G whose deletion destroys all (x, y)-paths in G.*

Proof   Construct a digraph $D(G)$ from G as follows: $V(G)$ is also the vertex set of $D(G)$ and if $u, v \in V(G)$, then $(u, v) \in A(D(G))$ if, and only if, u and v are adjacent in G, that is, $D(G)$ is obtained from G by replacing each edge uv of G by a symmetric pair of arcs (u, v) and (v, u). By Theorem 3.5.11, the maximum number of arc-disjoint directed (x, y)-paths in $D(G)$ is equal to the minimum number of arcs whose deletion destroys all directed (x, y)-paths in $D(G)$. But each directed (x, y)-path in $D(G)$ gives rise to a unique (x, y)-path in G, and, conversely, an (x, y)-path in G yields a unique directed (x, y)-path in $D(G)$. Hence, the deletion of a set of λ edges in G destroys all (x, y)-paths in G if, and only if, the deletion of the corresponding set of λ arcs in $D(G)$ destroys all directed (x, y)-paths in $D(G)$.   □

Theorem 3.5.13 is the *vertex version of Menger's theorem for digraphs.*

Theorem 3.5.13   *Let x and y be two vertices of a digraph D such that $(x, y) \notin A(D)$. Then the maximum number of internally-disjoint directed (x, y)-paths in D is equal to the minimum number of vertices whose deletion destroys all directed (x, y)-paths in D.*

Proof   Construct a new digraph D' from D as follows:
(a)   Split each vertex $v \in V \backslash \{x, y\}$ into two new vertices, v' and v'', and join them by an arc (v', v''), and
(b)   replace
   (i)   each arc (u, v) of D where $u \notin \{x, y\}$ and $v \notin \{x, y\}$ by the arc (u'', v'),
   (ii)   each arc (x, v) of D by (x, v') and (v, x) by (v'', x), and
   (iii)   each arc (v, y) of D by (v'', y) and (y, v) by (y, v').
   (See Figure 3.16.)

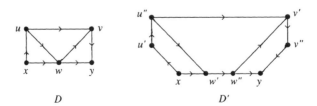

D D'

FIGURE 3.16. Digraphs D and D' for proof of Theorem 3.5.1

Now, to each directed (x, y)-path in D', there corresponds a directed (x, y)-path in D obtained by contracting all arcs of the type (v', v''), and, conversely, to each directed (x, y)-path in D, there corresponds a directed (x, y)-path in D' obtained by splitting each intermediate vertex of the path. Furthermore, two directed (x, y)-paths in D' are arc-disjoint if, and only if, the corresponding paths in D are internally disjoint. Hence, the maximum number of arc-disjoint directed (x, y)-paths in D' is equal to the maximum number of internally disjoint directed (x, y)-paths in D.

Similarly, the minimum number of arcs in D' whose deletion destroys all directed (x, y)-paths in D' is equal to the minimum number of vertices in D whose deletion destroys all directed (x, y)-paths in D. To see this, let A' be a minimum set of p arcs of D' whose deletion destroys all directed $x-y$ paths in D', and let V' be a minimum set of q vertices of D whose deletion destroys all directed $x-y$ paths in D. We have to show that $p = q$. Any arc of A' must contain either v' or v'', corresponding to some vertex v of D. Then the deletion of all vertices v corresponding to such arcs of D' separates x and y in D, and hence $q \leq p$. Conversely, if $v \in V'$, delete the corresponding arc (v', v'') in D'. Then the deletion of the q arcs (v', v'') (which correspond to the q vertices of V') from D' destroys all directed $x-y$ paths in D', and therefore $p \leq q$. Thus $p = q$. Now, the result follows from Theorem 3.5.11. □

Theorem 3.5.14 is the *vertex version of Menger's theorem for undirected graphs.*

Theorem 3.5.14   *Let x and y be two nonadjacent vertices of a graph G. Then the maximum number of internally disjoint (x, y)-paths in G is equal to the minimum number of vertices whose deletion destroys all (x, y)-paths.*

Proof   Define $D(G)$ as in Theorem 3.5.12 and apply Theorem 3.5.13. □

Let G be an undirected graph with $n \geq k + 1$ vertices. Suppose G satisfies the condition (*):

(*)   Any two distinct vertices of G are connected by k internally disjoint paths in G.

Then, Theorem 3.5.13 implies that to separate two nonadjacent vertices x and y of G, at least k vertices are to be removed. Hence if (*) holds, G is k-connected.

Conversely, if G is k-connected, to separate any pair of nonadjacent vertices x and y of G, at least k vertices are to be removed, and by Theorem 3.5.14, there are at least k internally disjoint $x-y$ paths in G. However, if x and y are adjacent, then since $G-xy$ is $(k - 1)$-connected, there are at least k internally disjoint $x-y$ paths, including the edge xy. Thus we have the following result of Whitney.

Theorem 3.5.15 (Whitney)   *A graph G with $n \geq k+1$ vertices is k-connected if, and only if, any two vertices of G are connected by at least k internally disjoint paths.*

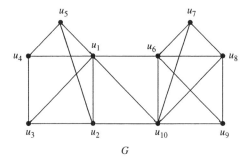

FIGURE 3.17. 2-connected and 3-edge-connected graph

Example 3.5.16   The graph G of Figure 3.17 is 2-connected and 3-edge connected. The pair of vertices u_5 and u_{10} are connected by the following two internally disjoint paths:

$$u_5 u_1 u_{10} \text{ and } u_5 u_4 u_3 u_2 u_{10}.$$

Moreover, they are connected by the following 3-edge-disjoint paths:

$$u_5 u_1 u_{10}; \quad u_5 u_2 u_{10}; \text{ and } u_5 u_4 u_1 u_6 u_{10}.$$

Exercise 5.1   If u and v are vertices of a graph G such that any two u–v paths in G have an internal common vertex, show that all the u–v paths in G have an internal common vertex.

Exercise 5.2   Show that if G is k-connected, then $G \vee K_1$ is $(k+1)$-connected.

Exercise 5.3   Let S be a subset of the vertex set of a k-connected graph G with $|S| = k$. If $v \in V \backslash S$, show that there exist k internally disjoint paths from v to the k vertices of S. (Remark: In particular, if C is a cycle of length at least k in G, and $v \in V(G) \backslash V(C)$, then there are k internally disjoint paths from v to C.)

The remark in Exercise 5.3 yields the following theorem of Dirac [33], which generalizes Theorem 3.2.8.

Exercise 5.4   *Dirac's Theorem [33]*: If a graph is k-connected ($k \geq 2$), then any set of k vertices of G lie on a cycle of G. (Note: The cycle may contain additional vertices besides these k vertices.) Hint: Use induction on k. If G is $(k+1)$-connected, and $\{v_1, v_2, \ldots, v_k, v_{k+1}\}$ is any set of $k+1$ vertices of G, by the induction assumption, $v_1, v_2, \ldots, v_k$ all lie on a cycle C of G. If $V(C) = \{v_1, v_2, \ldots, v_k\}$, then the k disjoint paths from v_{k+1} to C must end in $v_1, v_2, \ldots, v_k$. Otherwise, $V(C) \supsetneq \{v_1, v_2, \ldots, v_k\}$. Since G is $(k+1)$-connected, the end vertices of two of the $(k+1)$ disjoint paths from v_{k+1} to C must belong to one of the k closed paths $[v_i, v_{i+1}]$, $1 \leq i \leq k-1$ and $[v_k, v_1]$ on C. (Here $[v_i, v_{i+1}]$ and $[v_k, v_1]$ are those paths on C that contain no other v_j.)

FIGURE 3.18. A θ-graph

Exercise 5.5 Show by means of a simple example that the converse of Dirac's Theorem (Exercise 5.4) is false.

Exercise 5.6 Show that a k-connected simple graph on $(k+1)$ vertices is K_{k+1}.

Exercise 5.7 *Dirac's Theorem [33]; see also [61]:* Show that a graph G with at least $2k$ vertices is k-connected if, and only if, for any two disjoint sets V_1 and V_2 of k vertices each, there exist k disjoint paths from V_1 to V_2 in G.

Exercise 5.8 Show that a 2-connected non-Hamiltonian graph contains a θ-subgraph. (A θ-graph is a graph of the form $C \cup P$, where C is a cycle of length at least 4 and P is a path of length at least 2 that joins two nonadjacent vertices of C and is internally disjoint from C.) (See Figure 3.18.)

3.6. Exercises

6.1 Prove that there exists no simple connected cubic graph with less than 10 vertices containing a cut edge. Construct a simple connected cubic graph having exactly 10 vertices and having a cut edge.

6.2 Show that no vertex v of a simple graph can be a cut vertex of both G and G^c.

6.3 Show that if a simple connected graph is not a block, then it contains at least two end blocks.

6.4 Show that a connected k-regular bipartite graph is 2-connected.

6.5 Let $b(v)$ denote the number of blocks of a simple connected graph G to which a vertex v belongs. Then prove that the number of blocks $b(G)$ of G is given by $b(G) = 1 + \sum_{v \in V(G)} (b(v) - 1)$.
     (Hint: Use induction on the number of blocks of G.)

6.6 If $c(B)$ denotes the number of cut vertices of a simple connected graph G belonging to the block B, prove that the number of cut vertices $c(G)$ of G is given by $c(G) = 1 + \sum (c(B) - 1)$, the summation being over the blocks of G.

6.7 Show that a simple connected graph with at least three vertices is a path if, and only if, it has exactly two vertices that are not cut vertices.

6.8   Prove that if a graph G is k-connected or k-edge-connected, then $m \geq \frac{nk}{2}$.

6.9   Construct a graph with $\kappa = 3$, $\lambda = 4$, and $\delta = 5$.

6.10   For any three positive integers, a, b, c, with $a \leq b \leq c$, construct a simple graph with $\kappa = a$, $\lambda = b$, and $\delta = c$.

6.11   Let G be a cubic graph with a *1-factor* (i.e., a 1-regular spanning subgraph) F of G. Prove that any cut edge of G belongs to F.

6.12   Let G be a k-connected graph and let S be a separating set of G^2 such that $G^2 - S$ has q components. Show that $|S| \geq qk$.

6.13   Find all the edge cuts of the graph:

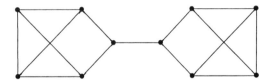

6.14   Let G be a 2-connected graph, and let $v_1, v_2 \in V(G)$. Let n_1 and n_2 be positive integers with $n = n_1 + n_2$. Then show that there exists a partition of V into $V_1 \cup V_2$ with $|V_i| = n_i$, $G[V_i]$ connected, and $v_i \in V_i$ for each $i = 1, 2$. (Remark: The generalization of this result to k-connected graphs is also true (see reference [33].)

Notes

Chronologically, Menger's Theorem appeared first [88]. Then followed Whitney's generalizations [121] of Menger's Theorem. Our proof of Menger's Theorem is based on the max-flow min-cut theorem of Ford and Fulkerson (references [42] and [43]).

IV

TREES

4.0. Introduction

"Trees" form an important class of graphs. Of late, their importance has grown considerably in view of their wide applicability in theoretical computer science.

In this chapter, we present the basic structural properties of trees, their centers and centroids. In addition, we present two interesting consequences of the Tutte-Nashwilliams theorem on the existence of k pairwise edge-disjoint spanning trees in a simple connected graph. We also present Cayley's formula for the number of spanning trees in the labeled complete graph K_n.

4.1. Definition, Characterization, and Simple Properties

Certain graphs derive their names from their diagrams. A "tree" is one such graph. Formally, a connected graph without cycles is defined as a *tree*. A graph without cycles is called an *acyclic graph* or a *forest*. So each component of a forest is a tree. A forest may consist of just a single tree! Figure 4.1 displays two pairs of isomorphic trees.

Remarks 4.1.1

1. It follows from the definition that a forest (and hence a tree) is a simple graph.

2. A subgraph of a tree is a forest, and a connected subgraph of a tree T is a *subtree* of T.

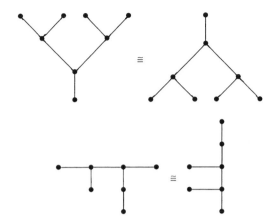

FIGURE 4.1. Examples of Isomorphic Trees

In a connected graph, any two distinct vertices are connected by at least one path. Trees are precisely those simple connected graphs in which every pair of distinct vertices is joined by a unique path.

Theorem 4.1.2   *A simple graph is a tree if, and only if, any two distinct vertices are connected by a unique path.*

Proof   Let T be a tree. Suppose that two distinct vertices u and v are connected by two distinct u–v paths. Then their union contains a cycle (cf. Exercise 4.9, Chapter I) in T, contradicting that T is a tree.

Conversely, suppose that any two vertices of a graph G are connected by a unique path. Then G is obviously connected. Also G cannot contain a cycle, since any two distinct vertices of a cycle are connected by two distinct paths. Hence G is a tree.   □

A spanning subgraph of a graph, which is also a tree, is called a *spanning tree* of the graph. A graph G and two of its spanning trees T_1 and T_2 are shown in Figure 4.2.

The graph G of Figure 4.2 shows that a graph may contain more than one spanning tree; each of the trees T_1 and T_2 is a spanning tree of G.

A loop cannot be an edge of any spanning tree, since such a loop constitutes a cycle (of length 1). On the other hand, a cut edge of G must be an edge of every spanning tree of G. Theorem 4.1.3 shows that every connected graph contains a spanning tree.

Theorem 4.1.3   *Every connected graph contains a spanning tree.*

Proof   Let G be a connected graph. Let $\mathscr{C}$ be the collection of all connected spanning subgraphs of G. $\mathscr{C}$ is nonempty as $G \in \mathscr{C}$. Let $T \in \mathscr{C}$ have the least

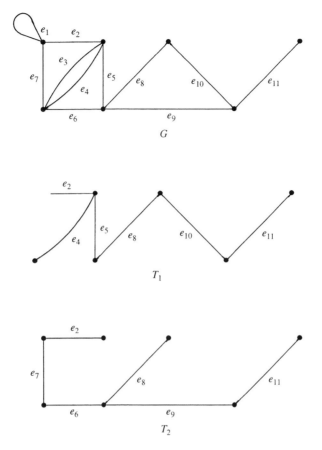

FIGURE 4.2. Graph G and its spanning trees T_1 and T_2

number of edges. Then T must be a spanning tree of G. If not, T would contain a cycle of G, and the deletion of any edge of this cycle would give a (spanning) subgraph in $\mathscr{C}$ having one edge less than that of T. This contradicts the choice of T. Hence T has no cycles and is therefore a spanning tree of G. □

There is a simple relation between the number of vertices and the number of edges of any tree.

Theorem 4.1.4 *The number of edges in a tree with n vertices is $n - 1$. Conversely, a connected graph with n vertices and $n - 1$ edges is a tree.*

Proof Let T be a tree. We now use induction on n to prove that $m = n - 1$. When $n = 1$ or $n = 2$, the result is straightforward.

Now assume that the result is true for all trees on $(n - 1)$ or fewer vertices, $n \geq 3$. Let T be a tree with n vertices. Let $e = uv$ be an edge of T. The path uv is the unique path in T joining u and v. Hence the deletion of e from T results in a disconnected graph having two components T_1 and T_2. Being connected subgraphs of a tree, T_1 and T_2 are themselves trees. As $n(T_1)$ and $n(T_2)$ are less than $n(T)$, by induction hypothesis, $m(T_1) = n(T_1) - 1$ and $m(T_2) = n(T_2) - 1$. Therefore $m(T) = m(T_1) + m(T_2) + 1 = n(T_1) - 1 + n(T_2) - 1 + 1 = n(T_1) + n(T_2) - 1 = n(T) - 1$.

Hence the result is true for T. By induction, the result follows in one direction.

Conversely, let G be a connected graph with n vertices and $n - 1$ edges. By Theorem 4.1.3, there exists a spanning tree T of G. T has n vertices, and, being a tree, has $(n - 1)$ edges. Hence $G = T$, and G is a tree. □

Exercise 1.1   Give an example of a graph with n vertices and $n - 1$ edges that is not a tree.

Theorem 4.1.5   *A tree with at least two vertices contains at least two pendant vertices (i.e., end vertices or vertices of degree 1).*

Proof   Consider a longest path of a tree T. The end vertices of this path must be pendant vertices of T; otherwise, the path is extendable to a longer path or else T contains a cycle, a contradiction. □

Corollary 4.1.6   *If $\delta(G) \geq 2$, then G contains a cycle.*

Proof   If G has no cycles, then G is a forest and hence $\delta(G) \leq 1$ by Theorem 4.1.5. □

Exercise 1.2   Show that a simple graph with ω components is a forest if, and only if, $m = n - \omega$.

Exercise 1.3   A vertex v of a tree T with at least three vertices is a cut vertex of T if, and only if, v is not a pendant vertex.

Exercise 1.4   Prove that every tree is a bipartite graph.

Our next result is a characterization of trees involving cut edges.

Theorem 4.1.7   *A connected graph G is a tree if, and only if, every edge of G is a cut edge of G.*

Proof   If G is a tree, there are no cycles in G. Hence no edge of G can belong to a cycle. By Theorem 3.1.7, each edge of G is a cut edge of G. Conversely, if every edge of a connected graph G is a cut edge of G, then G cannot contain a cycle, since the edges of a cycle are not cut edges of G. Hence G is a tree. □

Theorem 4.1.8 *A connected graph G with at least two vertices is a tree if, and only if, its degree sequence $(d_1, d_2, \ldots, d_n)$ satisfies the condition: $\sum_{i=1}^{n} d_i = 2(n-1)$ with $d_i > 0$ for each i.*

Proof Let G be a tree. As G is connected and nontrivial, it can have no isolated vertices. Hence every term of the degree sequence is positive. Further, by Theorem 1.3.4, $\sum_{i=1}^{n} d_i = 2m = 2(n-1)$.

Conversely, assume that the condition holds. We prove that G is a tree by induction on n. If $n = 2$, the condition $\sum_{i=1}^{n} d_i = 2(n-1)$ implies that $d_1 + d_2 = 2$. Since $d_1 \geq 1, d_2 \geq 1$, then $d_1 = 1, d_2 = 1$. The unique realization of G is a K_2, which is a tree. Now assume the result for all graphs with at most $(n-1)$ vertices $(n \geq 3)$. Let G have n vertices. If $d_i \geq 2$ for every i, $1 \leq i \leq n$, then $\sum_{i=1}^{n} d_i \geq 2n > 2(n-1)$, which contradicts the given condition. Hence $d_i < 2$ for at least one i. For this i, $d_i = 1$. For definiteness, let us assume that $d_n = 1$ and let v_n be the corresponding vertex. Let v_k be the unique vertex adjacent to v_n in G with $d_G(v_k) = d_k$. Delete v_n from the graph G. Then the resulting graph $G' = G - v_n$ is connected and has the degree sequence $(d_1, \ldots, d_{k-1}, d_k - 1, d_{k+1}, \ldots, d_{n-1})$. Let $v_1, \ldots, v_{n-1}$ be the respective vertices with these degrees. Note that d_k cannot be equal to 1, since in that case the edge $v_k v_n$ forms a separate component of G, contradicting that G is connected and $n \geq 3$. Thus, the degree sequence of G has positive terms. Further, $d_1 + \cdots + d_{k-1} + d_k - 1 + d_{k+1} + \cdots + d_{n-1} = (d_1 + \cdots + d_{n-1}) - 1 = (\sum_{i=1}^{n} d_i) - d_n - 1 = 2(n-1) - 1 - 1 = 2(n-2)$.

By induction hypothesis, G' is a tree. Now, to realize G from G', "attach" the pendant edge $v_k v_n$ at v_k. Clearly, G is also a tree. □

Lemma 4.1.9 *If u and v are nonadjacent vertices of a tree T, then $T + (uv)$ contains a unique cycle.*

Proof If P is the unique u–v path in T, then $P + (uv)$ is a cycle in $T + (uv)$. It is unique, as the path P is unique in T. □

Example 4.1.10 Prove that if $m(G) = n(G)$ for a simple connected graph G, then G is unicyclic, that is, a graph containing exactly one cycle.

Proof By Theorem 4.1.3, G contains a spanning tree. As T has $n(G) - 1$ edges, $E(G) \backslash E(T)$ consists of a single edge e. Then $G = T \cup e$ is unicyclic. □

Exercise 1.5 If for a simple graph G, $m(G) \geq n(G)$, prove that G contains a cycle.

Exercise 1.6 Prove that every edge of a connected graph G that is not a loop is in some spanning tree of G.

Exercise 1.7 Prove that the following statements are equivalent:

(i) G is connected and unicyclic (i.e., G has exactly one cycle).

(ii) G is connected and $n = m$.

(iii)   For some edge e of G, $G-e$ is a tree.

(iv) G is connected and the set of edges of G that are not cut edges forms a cycle.

Example 4.1.11   Prove that for a simple connected graph G, $L(G)$ is isomorphic to G if, and only if, G is a cycle.

Proof   If G is a cycle, then clearly $L(G)$ is isomorphic to G. Conversely, let $G \cong L(G)$. Then $n(G) = n(L(G))$, and $m(G) = m(L(G))$. But since $n(L(G)) = m(G)$, we have $m(G) = n(G)$. By Example 4.1.10, G is unicyclic. Let $C = v_1 v_2 \ldots v_k v_1$ be the unique cycle in G. If $G \neq C$, there must be an edge $e \notin E(C)$ incident with some vertex v_i of C (as G is connected). Thus, there is a star with at least three edges at v_i. This star induces a clique of size at least 3 in $L(G)$ ($\cong G$). This shows that there exists at least one cycle in $L(G)$ distinct from the cycle corresponding to C in G. This contradicts the fact that $L(G) \cong G$.   □

4.2.   Centers and Centroids

There are certain parameters attached to any connected graph. These are defined below.

Definitions 4.2.1   Let G be a connected graph.

1. The *diameter* of G is defined as $\max\{d(u, v) : u, v \in V(G)\}$ and is denoted by $\mathrm{diam}(G)$.

2. If v is a vertex of G, its *eccentricity* $e(v)$ is defined by $e(v) = \max\{d(u, v) : u \in V(G)\}$.

3. The *radius* of G, $r(G)$, is the minimum eccentricity of G, that is, $r(G) = \min\{e(v) : v \in V(G)\}$. Note that $\mathrm{diam}(G) = \max\{e(v) : v \in V(G)\}$.

4. A vertex v of G is called a *central vertex*, if $e(v) = r(G)$. The set of central vertices of G is called the *center* of G.

Example 4.2.2   Figure 4.3 displays two graphs T and G with the eccentricities of their vertices. We find that $r(T) = 4$ and $\mathrm{diam}(T) = 7$. Each of u and v is a central vertex of T. Also $r(G) = 3$, $\mathrm{diam}(G) = 4$. Further, G has five central vertices.

Remark 4.2.3   It is obvious that $r(G) \leq \mathrm{diam}(G)$. For a complete graph $r(G) = \mathrm{diam}(G) = 1$. For a complete bipartite graph $G(X, Y)$ with $|X| \geq 2$ and $|Y| \geq 2$, $r(G) = \mathrm{diam}(G) = 2$. For the graphs of Figure 4.3, $r(G) < \mathrm{diam}(G)$. The terms radius and diameter tempt one to expect that $\mathrm{diam}(G) = 2r(G)$. But this is

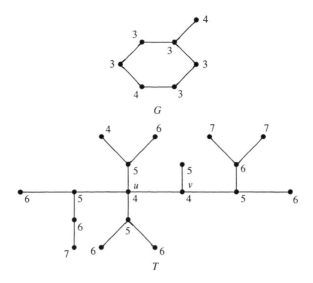

FIGURE 4.3. Eccentricities of vertices for graphs T and G

not the case, as the complete graphs show. In a tree, for any vertex u, $d(u, v)$ is maximum only when v is a pendant vertex. We use this observation in the proof of Theorem 4.2.4.

Theorem 4.2.4 (Jordan [73])   *Every tree has a center consisting of either a single vertex or two adjacent vertices.*

Proof   The result is obvious for the trees K_1 and K_2. The vertices of K_1 and K_2 are central vertices. Now let T be a tree with $n(T) \geq 3$. Then T has at least two pendant vertices (cf. Theorem 4.1.5). Clearly, the pendant vertices of T cannot be central vertices. Delete all the pendant vertices from T. This results in a subtree T' of T. As any maximum distance path in T from any vertex of T' ends at a pendant vertex of T, the eccentricity of each vertex of T' is one less than the eccentricity of the same vertex in T. Hence, the vertices of minimum eccentricity of T' are the same as those of T. In other words, T and T' have the same center. Now, if T'' is the tree obtained from T' by deleting all the pendant vertices of T', then T'' and T' have the same center. Hence the centers of T'' and T are the same. Repeat the process of deleting the pendant vertices in the successive subtrees of T until there results a K_1 or K_2. This will always be the case as T is finite. Hence, the center of T is either a single vertex or a pair of adjacent vertices. The process of determining the center described above is illustrated in Figure 4.4 for the tree T of Figure 4.3.

Exercise 2.1   Construct a tree with 85 vertices that has $\Delta = 5$ and the center consisting of a single vertex.

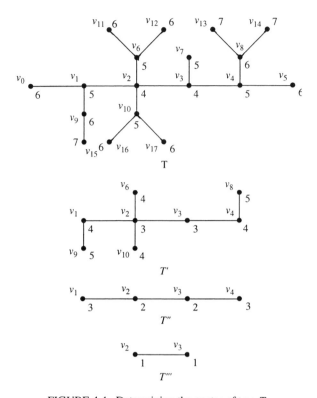

FIGURE 4.4. Determining the center of tree T

Exercise 2.2 Show that an automorphism of a tree on an odd number (≥ 3) of vertices has a fixed vertex, that is, for any automorphism f of a tree T with $n = 2k + 1$ ($k \geq 1$) vertices, there exists a vertex v of T with $f(v) = v$. (Hint: Use the fact that f permutes the end vertices of T.)

Exercise 2.3 Show that the distinct eccentricities of the vertices of a (connected) graph G form a set of consecutive integers starting from the radius of G and ending in the diameter of G.

Definitions 4.2.5

1. A *branch* at a vertex u of a tree T is a maximal subtree containing u as an end vertex. Hence, the number of branches at u is $d(u)$.

For instance, in Figure 4.5, there are three branches of the tree at u.

2. The *weight* of a vertex u of T is the maximum number of edges in any branch at u.

3. A vertex v is a *centroid vertex* of T if v has minimum weight. The set of all centroid vertices is called the *centroid* of T.

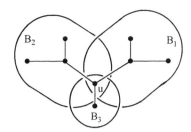

FIGURE 4.5. Tree showing three branches at u

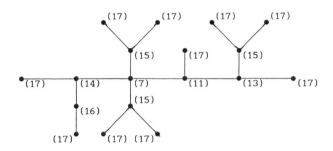

FIGURE 4.6. Weights of vertices of a tree

In Figure 4.6 the numbers in the parentheses indicate the weights of the corresponding vertices. It is clear that all the end vertices of T have the same weight, namely, $m(T)$.

As in the case of centers, any tree has a centroid consisting of either two adjacent vertices or a single vertex. But there is no relation between the center and centroid of a tree either with regard to the number of vertices or with regard to their location.

Exercise 2.4 Give an example of

(i) a tree with just one central vertex that is also a centroidal vertex;

(ii) a tree with two central vertices, one of which is also a centroidal vertex;

(iii) a tree with two centroidal vertices, one of which is also a central vertex;

(iv) a tree with two central vertices, both of which are also centroidal vertices; and

(v) a tree with disjoint center and centroid.

4.3.   Counting the Number of Spanning Trees

Counting the number of spanning trees in a graph occurs as a natural problem in many branches of science. Spanning trees were used by Kirchoff to generate a

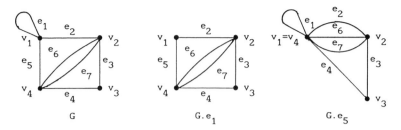

FIGURE 4.7. Edge contraction

"cycle basis" for the cycles in the graphs of electrical networks. In this section we consider enumeration of spanning trees of graphs.

The number of spanning trees of a connected labeled graph G will be denoted by $\tau(G)$. There is a recursive formula for $\tau(G)$. Before we establish this formula, we shall define the concept of *edge contraction* in graphs.

Definitions 4.3.1   An edge e of a graph G is said to be *contracted*, if it is deleted from G and its ends are identified. The resulting graph is denoted by $G.e$.

Edge contraction is illustrated in Figure 4.7.

If e is not a loop of G, then $n(G.e) = n(G) - 1, m(G.e) = m(G) - 1$, and $\omega(G.e) = \omega(G)$. For a loop $e, n(G.e) = n(G), m(G.e) = m(G) - 1$, and $\omega(G.e) = \omega(G)$. Theorem 4.3.2 gives a recursive formula for $\tau(G)$.

Theorem 4.3.2   *If e is not a loop of G, then $\tau(G) = \tau(G-e) + \tau(G.e)$.*

Proof   Let e be any edge of G that is not a loop. Then $\tau(G)$ is the sum of the number of spanning trees of G containing e and the number of spanning trees of G not containing e.

Since $V(G-e) = V(G)$, every spanning tree of $G-e$ is a spanning tree of G not containing e, and, conversely, any spanning tree of G for which e is not an edge is also a spanning tree of $G-e$. Hence the number of spanning trees of G not containing e is precisely the number of spanning trees of $G-e$, that is, $\tau(G-e)$.

If T is a spanning tree of G containing e, contraction of e in both T and G results in a spanning tree $T.e$ of $G.e$. Conversely, if T_0 is a spanning tree of $G.e$, there exists a unique spanning tree T of G containing e such that $T.e = T_0$. Thus, the number of spanning trees of G containing e is $\tau(G.e)$. Hence, $\tau(G) = \tau(G-e) + \tau(G.e)$. $\square$

We illustrate below the use of Theorem 4.3.2 in calculating the number of spanning trees. In this illustration, each graph within parentheses stands for the number of its spanning trees. For example, $[\boxdot]$ stands for the number of spanning trees of C_4.

Example 4.3.3   Find $\tau(G)$ for the following graph G:

G

Solution

$$\left[\;\text{🔲}\; \right]$$

$$= \left[\;\text{🔲}\; \right] + \left[\;\text{🔲}\; \right]$$

$$= \left[\;\text{🔲}\; \right] + \left[\;\text{🔲}\; \right]$$

$$= \left\{ \left[\;\text{🔲}\; \right] + \left[\;\text{🔲}\; \right] \right\} + \left[\;\text{🔲}\; \right]$$

$$= \left[\;\text{🔲}\; \right] + \left[\;\text{🔲}\; \right] + 2 \left[\;\text{🔲}\; \right]$$

$$= 1 + 3 + 2\,(4)$$

$$= 12$$

[By enumeration,

$$\left[\;\text{🔲}\; \right] = 1, \quad \left[\;\text{🔲}\; \right] = 3, \text{ and } \left[\;\text{🔲}\; \right] = 4. \,]$$

Hence $\tau(G) = 12$.   □

We have seen in Section 3.1 that every connected graph has a spanning tree. When will it have k edge-disjoint spanning trees? An answer to this interesting question was given by both Tutte [117] and Nash-Williams [94] at just about the same time.

Theorem 4.3.4 (Tutte [117] and Nash-Williams [94]) *A simple connected graph G contains k pairwise edge-disjoint spanning trees if, and only if, for each partition $\mathcal{P}$ of $V(G)$ into p parts, the number $m(\mathcal{P})$ of edges of G joining distinct parts is at least $k(p-1)$.*

Proof We prove only the easier part of the theorem. Suppose G has k pairwise edge-disjoint spanning trees. If T is one of them, and if $\mathcal{P} = \{V_1, \ldots, V_p\}$ is a partition of $V(G)$ into p parts, then identification of each part V_i into a single vertex v_i, $1 \le i \le p$, results in a connected graph G_0 (possibly with multiple edges) on $\{v_1, \ldots, v_p\}$. G_0 contains a spanning tree with $p-1$ edges, and each such edge belongs to T and joins distinct partite sets of $\mathcal{P}$. Since this is true for each of the k edge-disjoint spanning trees of G, the number of edges joining distinct parts of $\mathcal{P}$ is at least $k(p-1)$.

For the proof of the converse part of the theorem, we refer the reader to the references cited.

As a consequence of Theorem 4.3.4, we obtain immediately at least one family of graphs that possesses the property stated in the theorem.

Corollary 4.3.5 (Kundu [80]) *Every $2k$-edge-connected ($k \ge 1$) graph contains k pairwise edge-disjoint spanning trees.*

Proof Let G be $2k$-edge-connected, and let $\mathcal{P} = \{V_1, \ldots, V_p\}$ be a partition of V into p subsets. By hypothesis on G, there are at least $2k$ edges from each part V_i to $V \setminus V_i = \bigcup_{\substack{j=1 \\ j \ne i}}^{p} V_j$. The total number of such edges is $\ge kp$ (as each such edge is repeated twice). Hence $m(\mathcal{P}) \ge kp > k(p-1)$. Theorem 4.3.4 now ensures that there are at least k pairwise edge-disjoint spanning trees in G. □

Corollary 4.3.6 *Every 3-edge-connected graph has three spanning trees with empty intersection (i.e., a spanning totally disconnected subgraph).*

Proof Let G be a 3-edge connected graph. Duplicate each edge of G by a parallel edge. The resulting graph, say, G', is 6-edge-connected, and hence by Corollary 4.3.5, G' has three pairwise edge-disjoint spanning trees, say, T_1', T_2', and T_3'. Clearly, neither an edge of G nor its parallel edge can belong to all of T_1', T_2', and T_3', and hence $E(T_1' \cap T_2' \cap T_3') = \phi$. Let T_i, $1 \le i \le 3$ be the tree obtained from T_i' by replacing a parallel edge of G' by its original edge in G. Then, clearly, T_1, T_2, and T_3 are three spanning trees of G with $E(T_1 \cap T_2 \cap T_3) = \phi$. □

4.4. Cayley's Formula

Cayley was the first mathematician to obtain a formula for the number of spanning trees of a labeled complete graph.

Theorem 4.4.1 (Cayley [23]) $\tau(K_n) = n^{n-2}$ *where K_n is the labeled complete graph with n vertices, $n \geq 2$.*

Before we prove Theorem 4.4.1, we establish two lemmas.

Lemma 4.4.2   *Let $(d_1, \ldots, d_n)$ be a sequence of positive integers with $\sum_{i=1}^{n} d_i = 2(n-1)$, then there exists a tree T with vertex set $\{v_1, \ldots, v_n\}$ and $d(v_i) = d_i$, $1 \leq i \leq n$.*

Proof   It is easy to prove the result by induction on n.   □

Lemma 4.4.3   *Let $\{v_1, \ldots, v_n\}$, $n \geq 2$ be given and let $\{d_1, \ldots, d_n\}$ be a set of positive integers such that $\sum_{i=1}^{n} d_i = 2(n-1)$. Then the number of trees with $\{v_1, \ldots, v_n\}$ as the vertex set in which v_i has degree d_i, $1 \leq i \leq n$, is $\frac{(n-2)!}{(d_1-1)! \ldots (d_n-1)!}$.*

Proof   We prove the result by induction on n. For $n = 2$, $2(n-1) = 2$ so that $d_1 + d_2 = 2$. Since $d_1 \geq 1$ and $d_2 \geq 1$, $d_1 = d_2 = 1$. Hence, K_2 is the only tree in which v_i has degree d_i, $i = 1,2$. So the result is true for $n = 2$. Now assume that the result is true for all positive integers up to $n - 1$, $n \geq 3$. Let $d_1, \ldots, d_n$ be a sequence of positive integers such that $\sum_{i=1}^{n} d_i = 2(n-1)$ and let $\{v_1, \ldots, v_n\}$ be any set. If $d_i \geq 2$ for every i, $1 \leq i \leq n$, then $\sum_{i=1}^{n} d_i \geq 2n$. Hence, there exists an i, $1 \leq i \leq n$, for which $d_i = 1$. For the sake of definiteness, assume that $d_n = 1$. By Lemma 4.4.2, there exists a tree T with $V(T) = \{v_1, \ldots, v_n\}$ and degree of $v_i = d_i$. Let v_j be the unique vertex of T adjacent to v_n. Delete v_n from T. The resulting graph is a tree T' with $\{v_1, \ldots, v_{n-1}\}$ as its vertex set and $(d_1, \ldots, d_j - 1, \ldots, d_{n-1})$ as its degree sequence. In the opposite direction, given a tree T' with $\{v_1, \ldots, v_{n-1}\}$ as its vertex set and $(d_1, \ldots, d_j - 1, \ldots, d_{n-1})$ as its degree sequence, a tree T with vertex set $\{v_1, \ldots, v_n\}$ and degree sequence $(d_1, \ldots, d_n)$, $d_n = 1$, can be obtained by introducing a new vertex v_n and taking $T = T' + v_j v_n$. Hence, the number of spanning trees with vertex set $\{v_1, \ldots, v_n\}$ and degree sequence $(d_1, \ldots, d_n)$ with $d_n =$ degree of $v_n = 1$ and v_n adjacent to v_j is the same as the number of spanning trees with vertex set $\{v_1, \ldots, v_{n-1}\}$ and degree sequence $(d_1, \ldots, d_j - 1, \ldots, d_{n-1})$. By the induction hypothesis, the latter number is equal to

$$\frac{(n-3)!}{(d_1 - 1)! \ldots (d_{j-1} - 1)!(d_j - 2)!(d_{j+1} - 1)! \ldots (d_{n-1} - 1)!}$$

$$= \frac{(n-3)!(d_j - 1)}{(d_1 - 1)! \ldots (d_{j-1} - 1)!(d_j - 1)!(d_{j+1} - 1)! \ldots (d_{n-1} - 1)!}.$$

Summing over j, the number of trees with $\{v_1, \ldots, v_n\}$ as its vertex set and $(d_1, \ldots, d_n)$ as its degree sequence is

$$\sum_{j=1}^{n-1} \frac{(n-3)!(d_j - 1)}{(d_1 - 1)! \ldots (d_{n-1} - 1)!} = \frac{(n-3)!}{(d_1 - 1)! \ldots (d_{n-1} - 1)!} \sum_{j=1}^{n-1} (d_j - 1)$$

$$= \frac{(n-3)!}{(d_1 - 1)! \ldots (d_{n-1} - 1)!} \left[\left(\sum_{j=1}^{n-1} d_j \right) - (n-1) \right]$$

$$= \frac{(n-3)!}{(d_1 - 1)! \ldots (d_{n-1} - 1)!} [(2n-3) - (n-1)]$$

$$= \frac{(n-3)!}{(d_1 - 1)! \ldots (d_{n-1} - 1)!} (n-2)$$

$$= \frac{(n-2)!}{(d_1 - 1)! \ldots (d_{n-1} - 1)!}$$

$$= \frac{(n-2)!}{(d_1 - 1)! \ldots (d_n - 1)!} \quad \text{(recall that } d_n = 1\text{)}.$$

This completes the proof of Lemma 4.4.2. □

Proof of Theorem 4.4.1 The total number of trees T_n with vertex set $\{v_1, \ldots, v_n\}$ is obtained by summing over all possible sequences $(d_1, \ldots, d_n)$ with $\sum_{i=1}^{n} d_i = 2n - 2$. Hence,

$$\tau(K_n) = \sum_{d_i \geq 1} \frac{(n-2)!}{(d_1 - 1)! \ldots (d_n - 1)!} \quad \text{with} \quad \sum_{i=1}^{n} d_i = 2n - 2$$

$$= \sum_{k_i \geq 0} \frac{(n-2)!}{k_1! \ldots k_n!} \quad \text{with} \quad \sum_{i=1}^{n} k_i = n - 2, \text{ where } k_i = d_i - 1, \ 1 \leq i \leq n.$$

Putting $x_1 = x_2 = \cdots = x_n = 1$ and $m = n - 2$ in the multinomial expansion $(x_1 + x_2 + \cdots + x_n)^m = \sum_{k_i \geq 0} \frac{x_1^{k_1} x_2^{k_2} \ldots x_n^{k_n}}{k_1! k_2! \ldots k_n!} m!$ with $(k_1 + k_2 + \cdots + k_n) = m$, we get $n^{n-2} = \sum_{k_i \geq 0} \frac{(n-2)!}{k_1! k_2! \ldots k_n!}$ with $(k_1 + k_2 + \cdots + k_n) = n - 2$. Thus $\tau(K_n) = n^{n-2}$.

□

4.5. Helly Property

Definitions 4.5.1 A family $\{A_i : i \in I\}$ of subsets of a set A is said to satisfy the *Helly property* if $J \subseteq I$ and $A_i \cap A_j \neq \phi$ for every $i, j \in J$, then $\cap_{j \in J} A_j \neq \phi$.

Theorem 4.5.2 *A family of subtrees of a tree satisfies the Helly property.*

Proof Let $\mathscr{F} = \{T_i : i \in I\}$ be a family of subtrees of a tree T. Suppose that for all $i, j \in J \subseteq I$, $T_i \cap T_j \neq \phi$. We have to prove $\cap_{j \in J} T_j \neq \phi$. If some tree $T_i \in \mathscr{F}, i \in J$, is a single vertex tree $\{v\}$ (i.e., K_1), then, clearly, $\cap_{j \in J} T_j = \{v\}$. We therefore suppose that each tree $T_i \in \mathscr{F}$ with $i \in J$ has at least two vertices.

We now apply induction on the number of vertices of T. Let the result be true for all trees with at most n vertices, and let T be a tree with $(n + 1)$ vertices. Let v_0 be an end vertex of T, and u_0 its unique neighbor in T. Let $T_i' = T_i - v_0$, $i \in J$ and $T' = T - v_0$. By induction assumption, the result is true for the tree T'. Moreover, $T_i' \cap T_j' \neq \phi$ for any $i, j \in J$. In fact, if T_i and T_j have a vertex $u (\neq v_0)$ in common, then T_i' and T_j' also have u in common, whereas if T_i and T_j have v_0 in common, then T_i and T_j have u_0 also in common and so do T_i' and T_j'. Hence, by the induction assumption, $\cap_{j \in J} T_j' \neq \phi$, and therefore $\cap_{j \in J} T_j \neq \phi$. □

Exercise 5.1 In the cycle C_5, give a family of five paths such that the intersection of the vertex sets of any two of them is nonempty and the intersection of the vertex sets of all them is empty.

Exercise 5.2 Prove that a connected graph G is a tree if, and only if, every family of paths in G satisfies the Helly property.

4.6. Exercises

6.1 Show that any tree of order n contains a subtree of order m for every $m \leq n$.

6.2 Let u, v, w be any three vertices of a tree T. Show that either u, v, w all lie in a path of T or else there exists a vertex z of T which is common to the u–v, v–w, w–u paths of T.

6.3 Show that in a tree, the number of vertices of degree at least 3 is at most the number of end vertices minus 2.

6.4 Show that if G is a connected graph with at least three vertices, then G contains two vertices u and v such that G–$\{u, v\}$ is also connected.

6.5* If H is a graph of minimum degree at least $k - 1$, then prove that H contains every tree on k vertices. (Hint: Prove by induction on k.) (See reference [57].)

6.6 A set S of edges of a simple graph G is said to separate a set M of vertices, if no two vertices in M belong to the same component of G–S. Prove that a nontrivial simple graph G is a tree if, and only if, for any set of r distinct vertices in G, $r \geq 2$, the minimum number of edges required to separate them is $r - 1$. (See E. Sampathkumar [109].)

6.7 Show that a simple connected graph contains at least $m - n + 1$ distinct cycles.

6.8 Prove that, for a connected graph G, $r(G) \leq \text{diam}(G) \leq 2r(G)$. (The graphs of Figure 4.3 show that the inequalities can be strict.)

6.9 Prove that a tree with at least three vertices has diameter 2 if, and only if, it is a star.

6.10   Determine the number of spanning trees of the following graphs:

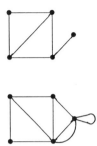

6.11   If T is a tree with at least two vertices, show that there exists a set of edge-disjoint paths covering all the vertices of T such that each of these paths has at least one end vertex that is an end vertex of T.

6.12   Let T be a tree of order n with $V(T) = \{1, 2, \ldots, n\}$, and let A be a set of transpositions defined by $A = \{(i, j) : ij \in E(T)\}$. Show that A is minimal set of transpositions that generates the symmetric group S_n.

Notes

Trees were first discovered by the English mathematician A. Cayley in 1857 when he wanted to enumerate the isomers of the saturated hydrocarbons $C_n H_{2n+2}$. (See also Chapter I.) Since then "trees" have grown both vertically and horizontally. They are also widely used nowadays in computer science.

There is also a simpler (?) proof of the converse part of Theorem 4.3.4 using matroid theory (see pp. 126–127 of reference [120]).

Corollary 4.3.6 is due to Kilpatrick [78] but the elegant proof given here is due to Jaeger [71]. The proof of Cayley's theorem presented here (Theorem 4.4.1) is based on Moon [90], which also contains nine other proofs. The book by J. P. Serre [110] entitled "Trees" is mainly concerned with the connection between trees and the group $SL_2(Q_p)$.

V

Independent Sets and Matchings

5.0. Introduction

Vertex-independent sets and vertex coverings as also edge-independent sets and edge coverings of graphs occur very naturally in many practical situations and hence have several potential applications. In this chapter, we study the properties of these sets. In addition, we also discuss matchings in graphs and, in particular, in bipartite graphs. Matchings in bipartite graphs have varied applications in operations research. We also present two celebrated theorems of graph theory, namely, Tutte's 1-factor theorem and Hall's matching theorem. **All graphs considered in this chapter are loopless**.

5.1. Vertex Independent Sets and Vertex Coverings

Definition 5.1.1 A subset S of the vertex set V of a graph G is called *independent* if no two vertices of S are adjacent in G. $S \subseteq V$ is a *maximum independent set* of G if G has no independent set S' with $|S'| > |S|$. A *maximal independent set* of G is an independent set that is not a proper subset of another independent set of G.

For example, in the graph of Figure 5.1, $\{u, v, w\}$ is a maximum independent set and $\{x, y\}$ is a maximal independent set that is not maximum.

Definition 5.1.2 A subset K of V is called a *covering* of G if every edge of G is incident with at least one vertex of K. A covering K is *minimum* if there is no

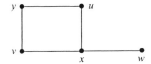

FIGURE 5.1. Graph with maximum independent set $\{u, v, w\}$ and maximal independent set $\{x, y\}$

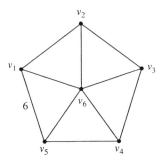

FIGURE 5.2. Wheel W_5

covering K' of G such that $|K'| < |K|$; it is *minimal* if there is no covering K_1 of G such that K_1 is a proper subset of K.

In the graph W_5 of Figure 5.2, $\{v_1, v_2, v_3, v_4, v_5\}$ is a covering of W_5 and $\{v_1, v_3, v_4, v_6\}$ is a minimal covering. Also, the set $\{x, y\}$ is a minimum covering of the graph of Figure 5.1.

The concepts of covering and independent sets of a graph arise very naturally in practical problems. Suppose we want to store a set of chemicals in different rooms. Naturally, we would like to store incompatible chemicals, that is, chemicals that are likely to react violently when brought together, in distinct rooms. Let G be a graph whose vertex set represents the set of chemicals and let two vertices be made adjacent in G if, and only if, the corresponding chemicals are incompatible. Then any set of vertices representing compatible chemicals forms an independent set of G.

Now consider the graph G whose vertices represent the various locations in a factory and whose edges represent the pathways between pairs of such locations. A light source placed at a location supplies light to all the pathways incident to that location. A set of light sources that supplies light to all the pathways in the factory forms a covering of G.

Theorem 5.1.3   *A subset S of V is independent if, and only if, $V \setminus S$ is a covering of G.*

Proof S is independent if, and only if, no two vertices in S are adjacent in G. Hence, every edge of G must be incident to a vertex of $V \setminus S$. This is the case if, and only if, $V \setminus S$ is a covering of G.   □

Definition 5.1.4   The number of vertices in a maximum independent set of G is called the *independence number* (or the *stability number*) of G and is denoted by $\alpha(G)$. The number of vertices in a minimum covering of G is the *covering number* of G and is denoted by $\beta(G)$

Corollary 5.1.5   *For any graph G, $\alpha + \beta = n$.*

Proof   Let S be a maximum independent set of G. By Theorem 5.1.3, $V \setminus S$ is a covering of G, and therefore $|V \setminus S| = n - \alpha \geq \beta$. Similarly, let K be a minimum covering of G. Then $V \setminus K$ is independent and so $|V \setminus K| = n - \beta \leq \alpha$. These two inequalities imply that $n = \alpha + \beta$.   □

5.2.   Edge-Independent Sets

Definitions 5.2.1
1. A subset M of the edge set E of a loopless graph G is called *independent* if no two edges of M are adjacent in G.
2. A *matching* in G is a set of independent edges.
3. An *edge covering* of G is a subset L of E such that every vertex of G is incident to some edge of L. Hence, an edge covering of G exists if, and only if, $\delta > 0$.
4. A matching M of G is *maximum* if G has no matching M' with $|M'| > |M|$. M is *maximal* if G has no matching M' strictly containing M. $\alpha'(G)$ is the cardinality of a maximum matching and $\beta'(G)$ is the size of a minimum edge covering of G.
5. A set S of vertices of G is said to be *saturated* by a matching M of S if every vertex of S is incident to some edge of M; a vertex v of G is *M-unsaturated* if it is not *M-saturated*.

For example, in the wheel W_5 (Figure 5.2), $M = \{v_1 v_2, v_4 v_6\}$ is a maximal matching; $\{v_1 v_5, v_2 v_3, v_4 v_6\}$ is a maximum matching and a minimum edge covering; the vertices $v_1, v_2, v_4,$ and v_6 are *M*-saturated, whereas v_3 and v_5 are *M*-unsaturated.

Remark 5.2.2   The edge analogue of Theorem 5.1.3, however, is not true. (For instance, in the graph G of Figure 5.3, the set $E' = \{e_3, e_4\}$ is independent and $E \setminus E' = \{e_1, e_2, e_5\}$ is not an edge covering of G. Also $E'' = \{e_1, e_3, e_4\}$ is an edge covering of G and $E \setminus E''$ is not independent in G. Again, E' is a matching in G that saturates $v_2, v_3, v_4,$ and v_5 but does not saturate v_1.

Theorem 5.2.3   *For any graph G for which $\delta > 0$, $\alpha' + \beta' = n$.*

Proof   Let M be a maximum matching in G so that $|M| = \alpha'$. Let U be the set of *M*-unsaturated vertices in G. Since M is maximum, U is an independent

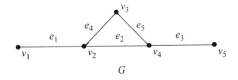

FIGURE 5.3. Graph illustrating edge relationships

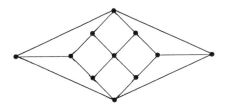

FIGURE 5.4. Herschel graph

set of vertices with $|U| = n - 2\alpha'$. Since $\delta > 0$, we can pick one edge, for each vertex in U, incident with it. Let F be the set of edges thus chosen. Then $M \cup F$ is an edge covering of G. Hence, $|M \cup F| = |M| + |F| = \alpha' + n - 2\alpha' \geq \beta'$, and therefore

$$n \geq \alpha' + \beta'. \tag{1}$$

Now let L be a minimum edge covering of G so that $|L| = \beta'$. Let $H = G[L]$, the edge subgraph of G defined by L, and let M_H be a maximum matching in H. Denote the set of M_H-unsaturated vertices in H by U. As L is an edge covering of G, H is a spanning subgraph of G. Consequently, $|L| - |M_H| = |L \setminus M_H| \geq |U| = n - 2|M_H|$ and so $|L| + |M_H| \geq n$. But since M_H is a matching in G, $|M_H| \leq \alpha'$. Thus,

$$n \leq |L| + |M_H| \leq \beta' + \alpha'. \tag{2}$$

Inequalities (1) and (2) imply that $\alpha' + \beta' = n$. □

Exercise 2.1 Determine the values of the parameters α, α', β, and β' for
(i) K_n,
(ii) the Petersen graph P,
(iii) the Herschel graph (see Figure 5.4).

Exercise 2.2 For any graph G with $\delta > 0$, prove that $\alpha \leq \beta'$ and $\alpha' \geq \beta$.

Exercise 2.3 Show that for a bipartite graph G, $\alpha\beta \geq m$ and that equality holds if, and only if, G is complete.

5.3.   Matchings and Factors

Definition 5.3.1   A *matching* of a graph G is (as given in Definition 5.2.1) a set of independent edges of G. If $e = uv$ is an edge of a matching M of G, the end vertices u and v of e are said to be *matched* by M.

If M_1 and M_2 are matchings of G, the edge subgraph defined by $M_1 \triangle M_2$, the *symmetric difference* of M_1 and M_2, is a subgraph H of G whose components are paths or even cycles of G in which the edges alternate between M_1 and M_2.

Definition 5.3.2   An *M-augmenting path* in G is a path in which the edges alternate between $E \backslash M$ and M and its end vertices are M-unsaturated. An *M-alternating path* in G is a path whose edges alternate between $E \backslash M$ and M.

Example 5.3.3   In the graph G of Figure 5.2, $M_1 = \{v_1v_2, v_3v_4, v_5v_6\}$ and $M_2 = \{v_1v_2, v_3v_6, v_4v_5\}$ and $M_3 = \{v_3v_4, v_5v_6\}$ are matchings of G. Moreover, $G[M_1 \triangle M_2]$ is the even cycle $v_3v_4v_5v_6v_3$. The path $v_2v_3v_4v_6v_5v_1$ is an M_3-augmenting path in G.

Maximum matchings have been characterized by Berge [13].

Theorem 5.3.4   *A matching M of a graph G is maximum if, and only if, G has no M-augmenting path.*

Proof   Assume first that M is maximum. If G has an M-augmenting path $P : v_0v_1v_2 \ldots v_{2t+1}$ in which the edges alternate between $E \backslash M$ and M, then P has one edge of $E \backslash M$ more than that of M. Define $M' = M \cup \{v_0v_1, v_2v_3, \ldots, v_{2t}v_{2t+1}\} \backslash \{v_1v_2, v_3v_4, \ldots, v_{2t-1}v_{2t}\}$. Clearly, M' is a matching of G with $|M'| = |M| + 1$, which is a contradiction, since M is a maximum matching of G.

Conversely, assume that G has no M-augmenting path. Then M must be maximum. If not, there exists a matching M' of G with $|M'| > |M|$. Let H be the edge subgraph $G[M \triangle M']$ defined by the symmetric difference of M and M'. Then the components of H are paths or even cycles in which the edges alternate between M and M'. Since $|M'| > |M|$, at least one of the components of H must start and end with edges of M'. But then such a path is an M-augmenting path of G, contradicting the assumption. See Figure 5.5. $\square$

Definition 5.3.5   A *factor* of a graph G is a spanning subgraph of G. A *k-factor* of G is a factor of G that is k-regular. Thus a *1-factor* of G is a matching that saturates all the vertices of G. For this reason, a 1-factor of G is called a *perfect matching* of G. A 2-factor of G is a factor of G that is a disjoint union of cycles of G. A graph G is *k-factorable* if G is an edge-disjoint union of k-factors of G.

Example 5.3.6   In Figure 5.6, G_1 is 1-factorable and G_2 is 2-factorable, whereas G_3 has neither a 1-factor nor a 2-factor. The dotted, solid, and parallel lines of G_1

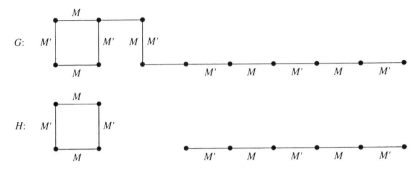

FIGURE 5.5. Graphs for proof of Theorem 5.3.4

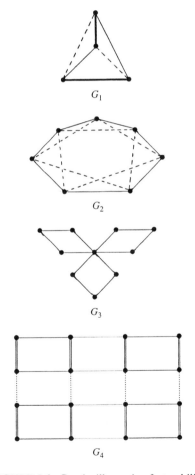

FIGURE 5.6. Graphs illustrating factorability

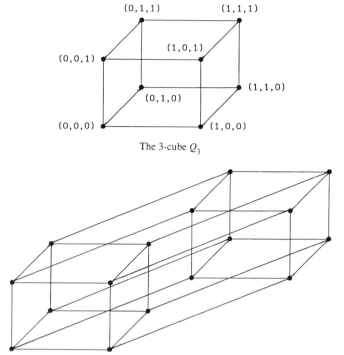

The 3-cube Q_3

The 4-cube Q_4

FIGURE 5.7. 3-cube (Q_3) and 4-cube (Q_4) graph

give the three distinct 1-factors, and the dotted and solid lines of G_2 give the two distinct 2-factors.

Exercise 3.1 Give an example of a cubic graph having no 1-factor.

Exercise 3.2 Show that $K_{n,n}$ and K_{2n} are 1-factorable.

Exercise 3.3 Show that the number of 1-factors of
(i) $K_{n,n}$ is $n!$,
(ii) K_{2n} is $\frac{(2n)!}{2^n n!}$.

Exercise 3.4 The *n-cube* Q_n is the graph whose vertices are binary n-tuples. Two vertices of Q_n are adjacent if, and only if, they differ in exactly one place. Show that $Q_n (n \geq 2)$ has a perfect matching. (The 3-cube Q_3 and the 4-cube Q_4 are displayed in Figure 5.7.)

Exercise 3.5 Show that the Petersen graph P is not 1-factorable. (Hint: Look at the possible types of 1-factors of P.)

Exercise 3.6 Show that every tree has at most one perfect matching.

Exercise 3.7* Show that if a 2-edge connected graph has a 1-factor, then it has at least two distinct 1-factors.

Exercise 3.8 Show that the graph G_4 of Figure 5.6 is not 1-factorable.

An Application to Physics 5.3.7 In crystal physics, a crystal is represented by a 3-dimensional lattice in which each face corresponds to a 2-dimensional lattice. Each vertex of the lattice represents an atom of the crystal, and an edge between two vertices represents the bond between the two corresponding atoms.

In crystallography, one is interested in obtaining an analytical expression for certain surface properties of crystals consisting of diatomic molecules (also called dimers). For this, one must find the number of ways in which all the atoms of the crystal can be paired off as molecules consisting of two atoms each. The problem is clearly equivalent to that of finding the number of perfect matchings of the corresponding 2-dimensional lattice.

Two different dimer coverings (perfect matchings) of the lattice defined by the graph G_4 are exhibited in Figure 5.6—one in solid lines and the other in parallel lines.

5.4. Matchings in Bipartite Graphs

Assignment Problem 5.4.1   Suppose in a factory there are n jobs, $j_1, j_2, \ldots, j_n$ and s workers, $w_1, w_2, \ldots, w_s$. Also suppose that each job j_i can be performed by a certain number of workers, and that each worker w_j has been trained to do a certain number of jobs. Is it possible to assign each of the n jobs to a worker who can do that job so that no two jobs are assigned to the same worker?

We convert this problem into a problem in graphs as follows: Form a bipartite graph G with bipartition (J, W), where $J = \{j_1, j_2, \ldots, j_n\}$ and $W = \{w_1, w_2, \ldots, w_s\}$ and make j_i adjacent to w_j if, and only if, worker w_j can do job j_i. Then our assignment problem translates into the following graph problem: Is it possible to find a matching in G that saturates all the vertices of J?

A solution to the above matching problem in bipartite graphs has been given by P. Hall [59] (see also M. Hall, Jr. [60]).

For a subset $S \subseteq V$ in a graph G, $N(S)$ denotes the neighbor set of S, that is, the set of all vertices each of which is adjacent to at least one vertex in S.

Theorem 5.4.2 (Hall)   *Let G be a bipartite graph with bipartition (X, Y). Then G has a matching that saturates all the vertices of X if, and only if,*

$$|N(S)| \geq |S| \qquad (*)$$

for every subset S of X.

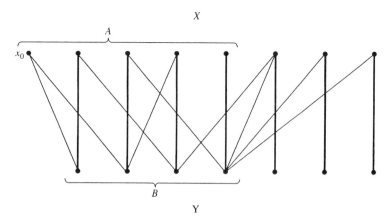

FIGURE 5.8. Figure for proof of Theorem 5.4.2 (matching edges are bold faced)

Proof   If G has a matching that saturates all the vertices of X, then each vertex of X is matched to a distinct vertex of Y. Hence, trivially, $|N(S)| \geq |S|$ for every subset $S \subseteq X$.

Conversely, assume that the condition $(*)$ above holds but that G has no matching that saturates all the vertices of X. Let M be a maximum matching of G. As M does not saturate all the vertices of X, there exists a vertex $x_0 \in X$ that is M-unsaturated. Let Z denote the set of all vertices of G connected to x_0 by M-alternating paths. Since M is a maximum matching, by Theorem 5.3.4, G has no M-augmenting path. As x_0 is M-unsaturated, this implies that x_0 is the only vertex of Z that is M-unsaturated. Let $A = Z \cap X$ and $B = Z \cap Y$. Then the vertices of $A \backslash \{x_0\}$ get matched under M to the vertices of B, and $N(A) = B$. Thus, since $|B| = |A| - 1$, $|N(A)| = |B| = |A| - 1 < |A|$, and this contradicts the assumption $(*)$. See Figure 5.8.   □

We now give some important consequences of Hall's Theorem.

Theorem 5.4.3   *A k (≥ 1)-regular bipartite graph is 1-factorable.*

Proof   Let G be k-regular with bipartition (X, Y). Then $E(G) =$ the set of edges incident to the vertices of $X =$ the set of edges incident to the vertices of Y. Hence, $k|X| = |E(G)| = k|Y|$ and therefore $|X| = |Y|$. If $S \subseteq X$, then $N(S) \subseteq Y$ and $N(N(S))$ contains S. Let E_1 and E_2 be the sets of edges of G incident to S and $N(S)$, respectively. Then $E_1 \subseteq E_2$, $|E_1| = k|S|$, and $|E_2| = k|N(S)|$. Hence, as $|E_2| \geq |E_1|$, $|N(S)| \geq |S|$. So by Hall's theorem (Theorem 5.4.2), G has a matching that saturates all the vertices of X; that is, G has a perfect matching M. Deletion of the edges of M from G results in a $(k - 1)$-regular bipartite graph. Repeated application of the above argument shows that G is 1-factorable.   □

König's Theorem: Consider any matching M of a graph G. If K is any (vertex) covering for the graph, then it is clear that to cover each edge of M we have to

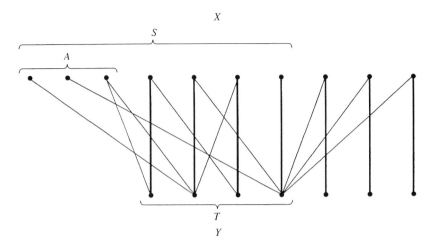

FIGURE 5.9. Graph for proof of Theorem 5.4.5

choose at least one vertex of K. Thus $|M| \leq |K|$. In particular, if M^* is a maximum matching and K^* is a minimum covering of G, then

$$|M^*| \leq |K^*| \qquad (**)$$

König's theorem asserts that for loopless bipartite graphs, equality holds in relation $(**)$. Before we establish this theorem, we present a lemma that is interesting in its own right and is similar to Lemma 3.6.8.

Lemma 5.4.4   *Let K be any covering and M any matching of a graph G with $|K| = |M|$. Then K is a minimum covering and M is a maximum matching.*

Proof   Let M^* be a maximum matching and K^* be a minimum covering of G. Then $|M| \leq |M^*|$ and $|K| \geq |K^*|$. Hence, by $(**)$ we have $|M| \leq |M^*| \leq |K^*| \leq |K|$. Since $|M| = |K|$, we must have $|M| = |M^*| = |K^*| = |K|$, proving the lemma.   □

Theorem 5.4.5 (König)   *In a loopless bipartite graph the minimum number of vertices that cover all the edges of G is equal to the maximum number of independent edges, that is, $\alpha' = \beta$.*

Proof   Let G be a bipartite graph with bipartition (X, Y). Let M be a maximum matching in G. Denote by A the set of vertices unsaturated by M (see Figure 5.9). As in the proof of Theorem 5.4.2, let Z stand for the set of vertices connected to A by M-alternating paths starting in A. Let $S = X \cap Z$ and $T = Y \cap Z$. Then clearly, $T = N(S)$ and $K = T \cup (X \backslash S)$ is a covering of G, because if there is an edge e not incident to any vertex in K, then one of the end vertices of e must be in S and the other in $Y \backslash T$. But this contradicts the fact that $N(S) = T$. Clearly, $|K| = |M|$, and so by Lemma 5.4.4, M is a maximum matching and K is a minimum covering of G.   □

Let A be a binary matrix (so that each entry of A is 0 or 1). A *line* of A is a row or column of A. A line covers all of its entries. Two 1's of A are called *independent* if they do not lie in the same line of A. Then the matrix version of König's theorem is given in Theorem 5.4.6.

Theorem 5.4.6 (Matrix version of König's theorem)   *In a binary matrix, the minimum number of lines that cover all the 1's is equal to the maximum number of independent 1's.*

Proof   Let $A = (a_{ij})$ be a binary matrix of size p by q. Form a bipartite graph G with bipartition (X, Y) where X and Y are sets of cardinality p and q, respectively, say, $X = \{v_1, v_2, \ldots, v_p\}$ and $Y = \{w_1, w_2, \ldots, w_q\}$. Make v_i adjacent to w_j in G if, and only if, $a_{ij} = 1$. Then an entry 1 in A corresponds to an edge of G, and two independent 1's correspond to two independent edges of G. Further, each vertex of G corresponds to a line of A. Thus the matrix version of König's theorem is actually a restatement of König's theorem.   □

A consequence of König's theorem is the theorem on the existence of a *System of Distinct Representatives* (SDR) for a family of subsets of a given finite set.

Definition 5.4.7   Let $\mathscr{F} = \{A_\alpha : \alpha \in J\}$ be a family of sets. An SDR for the family $\mathscr{F}$ is a family of elements $\{x_\alpha : \alpha \in J\}$ such that $x_\alpha \in A_\alpha$ for every $\alpha \in J$ and $x_\alpha \neq x_\beta$ whenever $\alpha \neq \beta$.

Example 5.4.8   For instance, if $A_1 = \{1\}$, $A_2 = \{2, 3\}$, $A_3 = \{3, 4\}$, $A_4 = \{1, 2, 3, 4\}$, and $A_5 = \{2, 3, 4\}$, then the family $\{A_1, A_2, A_3, A_4\}$ has $\{1, 2, 3, 4\}$ as an SDR, whereas the family $\{A_1, A_2, A_3, A_4, A_5\}$ has no SDR. It is clear that for $\mathscr{F}$ to have an SDR, it is necessary that for any positive integer k, the union of any k sets of $\mathscr{F}$ must contain at least k elements. That this condition is also sufficient when $\mathscr{F}$ is a finite family of finite sets is the assertion of Hall's theorem on the existence of an SDR.

Theorem 5.4.9 (Hall's theorem on the existence of an SDR [59])   *Let $\mathscr{F} = \{A_i : 1 \leq i \leq r\}$ be a family of finite sets. Then $\mathscr{F}$ has an SDR if, and only if, the union of any k, $1 \leq k \leq r$, members of $\mathscr{F}$ contains at least k elements.*

Proof   We need only prove the sufficiency part. Let $\bigcup_{i=1}^r A_i = \{y_1, y_2, \ldots, y_n\}$. Form a bipartite graph $G = G(X, Y)$ with $X = \{x_1, x_2, \ldots, x_r\}$, where x_i corresponds to the set A_i, $1 \leq i \leq r$, and $Y = \{y_1, y_2, \ldots, y_n\}$. Make x_i adjacent to y_j in G if, and only if, $y_j \in A_i$. Then it is clear that $\mathscr{F}$ has an SDR if, and only if, G

has a matching that saturates all the vertices of X. But this is the case, by Theorem 5.4.2, if for each $S \subseteq X$, $|N(S)| \geq |S|$, that is, if, and only if, $|\bigcup_{x_i \in S} A_i| \geq |S|$, which is precisely the condition stated in the theorem.   □

Exercise 4.1 Prove Theorem 5.4.5 (König's theorem) assuming Theorem 5.4.9.

Exercise 4.2 Show that a bipartite graph has a 1-factor if, and only if, $|N(S)| \geq |S|$ for *every* subset S of V. Does this hold for any graph G?

When does a graph have a 1-factor? Tutte's celebrated *1-factor theorem* answers this question. The proof given here is due to Lovász [85]. A component of a graph is *odd* or *even* according to whether it has an odd or even number of vertices. Let $O(G)$ denote the number of odd components of G.

Theorem 5.4.10 (Tutte's 1-factor theorem [116])   *A graph G has a 1-factor if, and only if,*

$$O(G-S) \leq |S|, \qquad\qquad (*)$$

for all $S \subseteq V$.

Proof While considering matchings in graphs, we are interested only in the adjacency of pairs of vertices. Hence, we may assume without loss of generality that G is simple. If G has a 1-factor M, then each of the odd components of $G-S$ must have at least one vertex, which is to be matched only to a vertex of S under M. Hence, for each odd component of $G-S$, there exists an edge of the matching with one end in S. Hence, the number of vertices in S should be at least as large as the number of odd components in $G-S$, that is, $O(G-S) \leq |S|$.

Conversely, assume that condition $(*)$ holds. If G has no 1-factor, we join pairs of nonadjacent vertices of G until we get a maximal supergraph G^* of G, with G^* having no 1-factor. Condition $(*)$ holds clearly for G^*, since

$$O(G^*-S) \leq O(G-S). \qquad\qquad (**)$$

(When two odd components are joined by an edge, there results an even component.)

Taking $S = \phi$ in $(*)$, we see that $O(G^*) = 0$, and so $n(G^*)\ (= n(G)) = n$ is even. Further, for every pair of nonadjacent vertices u and v of G^*, $G^* + uv$ has a 1-factor, and any such 1-factor must necessarily contain the edge uv.

Let K be the set of vertices of G^* of degree $(n-1)$. $K \neq V$, since otherwise $G^* = K_n$ has a perfect matching. We claim that each component of $G^* - K$ is complete. Suppose, to the contrary, that some component G_1 of $G^* - K$ is not complete. Then, in G_1 there are vertices x, y, and z such that $xy \in E(G^*)$, $yz \in E(G^*)$, but xz does not belong to $E(G^*)$ (Exercise 4.11 of Chapter 1). Moreover, since $y \in V(G_1)$, $d_{G^*}(y) < n - 1$ and hence there exists a vertex w of G^* with $yw \notin E(G^*)$. Necessarily, w does not belong to K. (See Figure 5.10.)

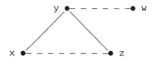

FIGURE 5.10. Supergraph G^* for proof of Theorem 5.4.10. Unbroken lines correspond to edges of G^* and broken lines correspond to edges not belonging to G^*

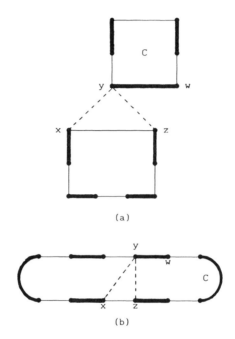

(a)

(b)

FIGURE 5.11. 1-factors M_1 and M_2 for (a) case 1 and (b) case 2 in proof of Theorem 5.4.10. Ordinary lines correspond to edges of M_1 and bold lines correspond to edges of M_2

By the choice of G^*, $G^* + xz$ and $G^* + yw$ have 1-factors, say M_1 and M_2, respectively. Necessarily, $xz \in M_1$ and $yw \in M_2$. Let H be the subgraph of $G^* + \{xz, yw\}$ induced by the edges in the symmetric difference $M_1 \triangle M_2$ of M_1 and M_2. Since M_1 and M_2 are 1-factors, each vertex of G^* is saturated by both M_1 and M_2, and H is a disjoint union of even cycles in which the edges alternate between M_1 and M_2. There arise two cases:

Case 1. xz and yw belong to different components of H (Figure 5.11(a)). If yw belongs to the cycle C, then the edges of M_1 in C together with the edges of M_2 not belonging to C form a 1-factor in G^*, contradicting the choice of G^*.

Case 2. xz and yw belong to the same component C of H. Since each component of H is a cycle, C is a cycle (Figure 5.11(b)). By symmetry of x and z, we may suppose that the vertices x, y, w, and z occur in that order on C. Then the edges of M_1 belonging to the $yw \cdots z$ section of C, together with the edge yz and the edges

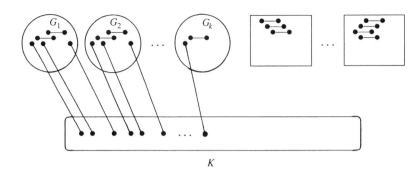

FIGURE 5.12. Components of $G^* - K$ for proof of Theorem 5.4.10

of M_2 not in the $yw \ldots z$ section of C form a 1-factor of G^*, again contradicting the choice of G^*. Thus each component of $G^* - K$ is complete.

By condition $(**)$, $O(G^* - K) \leq |K|$. Hence, one vertex of each of the odd components of $G^* - K$ is matched to a vertex of K. (This is possible, since each vertex of K is adjacent to every other vertex of G^*.) Also, the remaining vertices in each of the odd and even components of $G^* - K$ can be matched amongst themselves (see Figure 5.12). The total number of vertices thus matched is even. Since $|V(G^*)|$ is even, the remaining vertices of K can be matched amongst themselves. This gives a 1-factor of G^*. But, by choice, G^* has no 1-factor. This contradiction proves that G has a 1-factor. $\quad\square$

Corollary 5.4.11 (Petersen [104]) *Every 3-regular graph having no cut edges has a 1-factor.*

Proof Let G be a 3-regular graph without cut edges. Let $S \subseteq V$. Denote by $G_1, G_2, \ldots, G_k$ the odd components of $G-S$. Let m_i be the number of edges of G having one end in $V(G_i)$ and the other end in S. Since G is a cubic graph,

$$\sum_{v \in V(G_i)} d(v) = 3n(G_i) \qquad (1)$$

and

$$\sum_{v \in S} d(v) = 3|S|. \qquad (2)$$

Now, $E(G_i) = [V(G_i), V(G_i) \cup S] \setminus [V(G_i), S]$, where $[A, B]$ denotes the set of edges having one end in A and the other end in B, $A \subseteq V$, $B \subseteq V$. Hence, $m_i = |[V(G_i), S]| = \sum_{v \in V(G_i)} d(v) - 2m(G_i)$, and, since $d(v)$ is 3 for each v and $V(G_i)$ is an odd component, m_i is odd for each i. Further, as G has no cut edges, $m_i \geq 3$. Thus $O(G-S) = k \leq \frac{1}{3} \sum_{i=1}^{k} m_i \leq \frac{1}{3} \sum_{v \in S} d(v) = \frac{1}{3} 3|S| = |S|$. Therefore, by Tutte's theorem (Theorem 5.4.10), G has a 1-factor. $\quad\square$

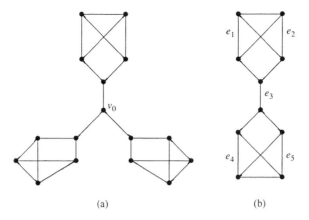

(a)   (b)

FIGURE 5.13. (a) 3-regular graph with cut edges having no 1-factor; (b) cubic graph with 1-factor having a cut edge

Example 5.4.12 A 3-regular graph with cut edges may not have a 1-factor. See Figure 5.13 (a). Again a cubic graph with a 1-factor may have cut edges. See Figure 5.13 (b).

In Figure 5.13 (a), if $S = \{v_0\}$, $O(G{-}S) = 3 > 1 = |S|$, and so G has no 1-factor. In Figure 5.13 (b), $\{e_1, e_2, e_3, e_4, e_5\}$ is a 1-factor, and e_3 is a cut edge of G. □

If G has no 1-factor, then by Theorem 5.4.10 there exists $S \subseteq V(G)$ with $O(G{-}S) > |S|$. Such a set S is called an *antifactor* set of G.

Let G be a graph of even order n and let S be an antifactor set of G. Let $O(G{-}S) = k$ and $G_1, G_2, \ldots, G_k$ be the odd components of $G{-}S$. Since n is even, $|S|$ and k have the same parity. Choose a vertex $u_i \in V(G_i)$, $1 \le i \le k$. Then $|S \cup \{u_1, u_2, \ldots, u_k\}|$ is even; that is, $k + |S| \equiv 0 \pmod 2$. This means that $k \equiv |S| \pmod 2$, and thus $O(G{-}S) \equiv |S| \pmod 2$. Thus we make the following observation.

Observation 5.4.13 If S is an antifactor set of a graph G of even order, then $O(G{-}S) \ge |S| + 2$.

Corollary 5.4.14 (W. H. Cunnigham; see reference [75]) *The edge set of a simple 2-edge-connected cubic graph G can be partitioned into paths of length three.*

Proof By Corollary 5.4.11, G is a union of a 1-factor and a 2-factor. Orient the edges of each cycle of the above 2-factor in any manner so that each cycle becomes a directed cycle. Then if e is any edge of the 1-factor, and f_1, f_2 are the two arcs of G having their tails at the end vertices of e, then $\{e, f_1, f_2\}$ forms a typical 3-path of the edge partition of G (see Figure 5.14). □

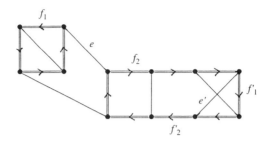

FIGURE 5.14. Figure for the proof of Corollary 5.4.14

Corollary 5.4.15   *A $(p-1)$-regular simple graph on $2p$ vertices has a 1-factor.*

Proof   Proof is by contradiction. Let G be a $(p-1)$-regular simple graph on $2p$ vertices having no 1-factor. Then G has an antifactor set S. By Observation 5.4.13, $O(G-S) \geq |S| + 2$. Hence $|S| + (|S| + 2) \leq 2p$, and therefore $|S| \leq p - 1$. Let $|S| = p - r$. Then $r \neq 1$ since if $r = 1$, $|S| = p - 1$, and therefore $O(G-S) = p + 1$. Hence, each odd component of $G-S$ is a singleton, and hence, each such vertex must be adjacent to all the $p - 1$ vertices of S (as G is $(p-1)$-regular). But this means that every vertex of S is of degree at least $p + 1$, a contradiction. Hence $|S| = p - r, 2 \leq r \leq p - 1$. If G' is any component of $G-S$ and $v \in V(G')$, then v can be adjacent to at most $|S|$ vertices of S. Therefore, as G is $(p-1)$-regular, v must be adjacent to at least $(p - 1) - (p - r) = r - 1$ vertices of G'. Thus $|V(G')| \geq r$. Counting the vertices of all the odd components of $G-S$ and the vertices of S, we get $(|S| + 2)r + |S| \leq 2p$, or $(p - r + 2)r + (p - r) \leq 2p$. This gives $(r - 1)(r - p) \geq 0$, violating the condition on r.   □

Our next result shows that there is a certain special family of graphs for which we can immediately conclude that all the graphs of the family have a 1-factor.

Theorem 5.4.16* (D. P. Sumner [111])   *Let G be a connected graph of even order n. If G is claw-free (i.e., contains no $K_{1,3}$ as an induced subgraph), then G has a 1-factor.*

Proof   If G has no 1-factor, G contains a minimal antifactor set S of G. There must be an edge between S and each odd component of $G-S$. If $v \in S$, and vx, vy, and vz are edges of G with x, y, and z belonging to distinct odd components of $G-S$, then a $K_{1,3}$ is induced in G. By hypothesis, this is not possible.

Since $O(G-S) > |S|$, there must certainly exist a vertex v of S, and edges vu and vw of G, with u and w in distinct odd components of $G-S$. Suppose G_u and G_w are the odd components containing u and w, respectively. Then $\langle G_u \cup G_w \cup \{v\} \rangle$ is an odd component of $G-S_1$, where $S_1 = S - \{v\}$. Further, $O(G-S_1) = O(G-S) - 1 > |S| - 1 = |S_1|$, and hence S_1 is an antifactor set of G with $|S_1| = |S| - 1$, a contradiction to the choice of S. Thus G must have a 1-factor. (Note that by Observation 5.4.13, the case $|S| = 1$ and $O(G-S) = 2$ cannot arise.)   □

Exercise 4.3 Find a 1-factorization of (i) Q_3, (ii) Q_4.

Exercise 4.4 Prove that Q_n, $n \geq 2$, is 1-factorable.

Exercise 4.5 Display a 2-factorization of K_9.

Exercise 4.6 Show that a k-regular $(k - 1)$-edge connected graph of even order has a 1-factor. (This result of F. Babler generalizes Petersen's result (Corollary 5.4.11) and can be shown by imitating its proof.)

Exercise 4.7 If G is a k-connected graph of even order having no $K_{1,k+1}$ as an induced subgraph, then show that G has a 1-factor.

Exercise 4.8 Show that if G is a connected graph of even order, then G^2 has a 1-factor.

5.5.* Perfect Matchings and the Tutte Matrix

It has been established by Tutte that the existence of a perfect matching in a simple graph is related to the nonsingularity of a certain square matrix. This matrix is called the "Tutte matrix" of the graph. We now define the Tutte matrix.

Definition 5.5.1 Let $G = (V, E)$ be a simple graph of order n and let $V = \{v_1, v_2, \ldots, v_n\}$. Let $\{x_{ij} : 1 \leq i < j \leq n\}$ be a set of indeterminates. Then the *Tutte matrix* of G is defined to be the n by n matrix $T = (t_{ij})$, where

$$t_{ij} = \begin{cases} x_{ij} & \text{if } v_i v_j \in E(G) \text{ and } i < j \\ -x_{ij} & \text{if } v_i v_j \in E(G) \text{ and } i > j \\ 0 & \text{otherwise.} \end{cases}$$

Thus T is a skew-symmetric matrix of order n.

Example 5.5.2 For example, if G is the graph

G

then

$$T = \begin{bmatrix} t_{11} & t_{12} & t_{13} & t_{14} \\ t_{21} & t_{22} & t_{23} & t_{24} \\ t_{31} & t_{32} & t_{33} & t_{34} \\ t_{41} & t_{42} & t_{43} & t_{44} \end{bmatrix} = \begin{bmatrix} 0 & 0 & x_{13} & x_{14} \\ 0 & 0 & x_{23} & x_{24} \\ -x_{13} & -x_{23} & 0 & x_{34} \\ -x_{14} & -x_{24} & -x_{34} & 0 \end{bmatrix}. \qquad (A)$$

Now, by the definition of a determinant of a square matrix, the determinant of $T(= \det T)$ is given by $\det T = \sum_{\pi \in S_n} \text{sgn}(\pi) t_{1\pi(1)} t_{2\pi(2)} \cdots t_{n\pi(n)}$, where $\pi \in S_n$ (i.e., π is a permutation on $\{1, 2, \ldots, n\}$), and $\text{sgn}(\pi) = 1$ or -1, according to whether π is an even or odd permutation, respectively. We denote the expression $t_{1\pi(1)} t_{2\pi(2)} \cdots t_{n\pi(n)}$ by t_π. Hence $\det T = \sum_{\pi \in S_n} \text{sgn}(\pi) t_\pi$. Further, if n is odd, say, $\pi = (123)$, then $t_\pi = t_{12} t_{23} t_{31} = x_{12} x_{23}(-x_{13})$ (Note: We take $x_{ij} = 0$ if $v_i v_j \notin E(G)$). Also $\pi^{-1} = (321)$ and $t_{\pi^{-1}} = t_{13} t_{21} t_{32} = x_{13}(-x_{12})(-x_{23})$, so that $t_\pi + t_{\pi^{-1}} = 0$. It is clear that the same thing is true for any odd $n \geq 3$.)

Now, for the Tutte matrix in relation (A), we have

$$\det T = x_{13}^2 x_{24}^2 + x_{14}^2 x_{23}^2 - 2 x_{13} x_{24} x_{14} x_{23}.$$

In this expression, the term $x_{13}^2 x_{24}^2$ is obtained by choosing the entries x_{13}, x_{24}, $-x_{13} = x_{31}$, and $-x_{24} = x_{42}$ of T, and hence it corresponds to the 1-factor $\{v_1 v_3, v_2 v_4\}$. Similarly, the term $x_{14}^2 x_{23}^2$ corresponds to the 1-factor $\{v_1 v_4, v_2 v_3\}$, and the term $x_{13} x_{24} x_{14} x_{23}$ corresponds to the cycle $(v_1 v_3 v_2 v_4)$ consisting of the edges $v_1 v_3$, $v_3 v_2$, $v_2 v_4$, and $v_4 v_1$.

We are now ready to prove Tutte's theorem, but before doing so, we make two useful observations.

Observation 5.5.3 If $\pi \in S_n$ is a product of disjoint even cycles, then $\text{sgn}(\pi) t_\pi = $ a product of squares of the form x_{ij}^2.

Indeed, in this case, n is even, and the edges of G corresponding to the alternate transpositions in each of the even cycles of π form a 1-factor of G. (For example, for the even cycle (1234), we take the alternate transpositions (12) and (34).) Further, if $v_i v_j$ $(i < j)$ is an edge of this 1-factor, the partial product $t_{ij} t_{ji} = -x_{ij}^2$ occurs in t_π. The number of such products is $\frac{n}{2}$, and therefore

$$\text{sgn}(\pi) t_\pi = (-1)^{\frac{n}{2}} (-1)^{\frac{n}{2}} \Pi x_{ij}^2 = \Pi x_{ij}^2,$$

where the product runs over all pairs (i, j) with $i < j$ such that $v_i v_j$ is an edge of the 1-factor corresponding to π.

Observation 5.5.4 If $\pi \in S_n$ has an odd cycle α in its decomposition into the product of disjoint cycles, consider $\pi_1 \in S_n$, where π_1 is obtained from π by replacing α by α^{-1} and retaining the remaining cycles in π. Then, from our earlier remarks, it is clear that $\text{sgn}(\pi) t_\pi + \text{sgn}(\pi_1) t_{\pi_1} = 0$.

Theorem 5.5.5 (W. T. Tutte) *Let G be a simple graph with Tutte matrix T. Then G has a 1-factor if, and only if, $\det T \neq 0$.*

Proof Suppose that $\det T \neq 0$. Then by Observation 5.5.3, there exists a $\pi \in S_n$ containing no odd cycle in its cycle decomposition. Then π is a product of even cycles and, by Observation 5.5.3, $\text{sgn}(\pi) t_\pi = \Pi x_{ij}^2$. The alternate transpositions of the even cycles of π then yield a 1-factor of G.

Conversely, assume that G has a 1-factor. Let $\pi \in S_n$ be the product of those transpositions corresponding to the 1-factor of G. (If $v_i v_j$ is an edge of the 1-factor, the corresponding transposition is (ij).) Then by Observation 5.5.3 $\mathrm{sgn}(\pi)t_\pi = \Pi x_{ij}^2$. Now set

$$x_{ij} = \begin{cases} 1 & \text{if } x_{ij}^2 \text{ appears in the product for } \mathrm{sgn}(\pi)t_\pi \\ 0 & \text{otherwise.} \end{cases}$$

Then $\mathrm{sgn}(\pi)t_\pi = 1$, and for these values of x_{ij}, $\mathrm{sgn}(\sigma)t_\sigma = 0$ for any $\sigma \in S_n, \sigma \neq \pi$. This means that the polynomial det T is not the zero polynomial. $\square$

Remark 5.5.6 Actually, our definition of the Tutte matrix of G depends on the order of the vertices of G. That is to say, the definition of T is based on regarding G as a labeled graph. However, if T is nonsingular with regard to one labeling of G, then the Tutte matrix of G will remain nonsingular with regard to any other labeling of G. This is because if T and T' are the Tutte matrices of G with regard to two labelings of G, then $T' = PTP^{-1}$, where P is a permutation matrix of order n. Hence, T is nonsingular if, and only if, T' is nonsingular.

Exercise 5.1 By evaluating the Tutte matrix of the following graph G, show that G has a 1-factor.

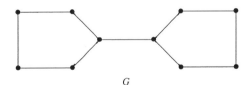

G

Notes

The reader who is more interested in matching theory can consult references [60], [86], and [103]. Our proof of Tutte's 1-factor theorem is due to Lovász [85] (see also [19]).

VI

EULERIAN AND HAMILTONIAN GRAPHS

6.0. Introduction

The study of Eulerian graphs was initiated in the 18th century, and that of Hamiltonian graphs in the 19th century. These graphs possess rich structure, and hence their study is a very fertile field of research for graph theorists. In this chapter, we present several structure theorems for these graphs.

6.1. Eulerian Graphs

Definition 6.1.1 An *Euler trail* in a graph G is a spanning trail in G that contains all the edges of G. An *Euler tour* of G is a closed Euler trail of G. G is called *Eulerian* (Figure 6.1(a)) if G has an *Euler tour*. It was Euler who first considered these graphs and hence their name.

It is clear that an Euler tour of G, if it exists, can be described from any vertex of G. Clearly, every Eulerian graph is connected.

It was Euler who showed in 1736 that the celebrated *Königsberg bridge problem* has no solution. The city of Königsberg (now called Kaliningrad) has seven bridges linking two islands A and B and the banks C and D of the Pregel (now called Pregalya) River as shown in Figure 6.2.

The problem was to start from any one of the four land areas, take a stroll across the seven bridges and get back to the starting point without crossing any bridge a second time. This problem can be converted into one concerning the

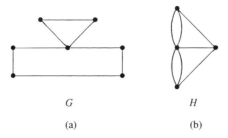

G   H

(a)   (b)

FIGURE 6.1. (a) Eurlerian graph G; (b) non-Eulerian graph H

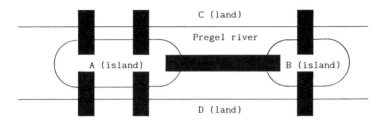

FIGURE 6.2. Königsberg bridge problem

graph obtained by representing each land area by a vertex and each bridge by an edge. The resulting graph H is the graph of Figure 6.1(b). The Königsberg bridge problem will have a solution provided that this graph H is Eulerian. But this is not the case, since it has vertices of odd degrees (see Theorem 6.1.2).

Eulerian graphs admit, among others, the following two elegant characterizations, Theorems 6.1.2 and 6.1.3.

Theorem 6.1.2 *For a connected graph G, the following statements are equivalent:*
   (i)   *G is Eulerian.*
  (ii)   *The degree of each vertex of G is an even positive integer.*
 (iii)   *G is an edge-disjoint union of cycles.*

Proof (i) $\Rightarrow$ (ii): Let T be an Euler tour of G described from some vertex $v_0 \in V(G)$. If $v \in V(G)$, and $v \neq v_0$, then every time T enters v, it must move out of v to get back to v_0. Hence, two edges incident with v are used during a visit to v, and hence, $d(v)$ is even. At v_0, every time T moves out of v_0, it must get back to v_0. Hence, $d(v_0)$ is also even. Thus the degree of each vertex of G is even.

  (ii) $\Rightarrow$ (iii): As $\delta(G) \geq 2$, G contains a cycle C_1 (Exercise 9.11 of Chapter I). In $G \setminus E(C_1)$, remove the isolated vertices, if there are any. Let the resulting subgraph of G be G_1. If G_1 is nonempty, each vertex of G_1 is again of even positive degree. Hence $\delta(G_1) \geq 2$, and so G_1 contains a cycle C_2. It follows that after a finite number, say r, of steps, $G \backslash E(C_1 \cup \cdots \cup C_r)$ is totally disconnected. Then G is the edge-disjoint union of the cycles $C_1, C_2, \ldots, C_r$.

(iii) $\Rightarrow$ (i): Assume that G is an edge-disjoint union of cycles. Since any cycle is Eulerian, G certainly contains an Eulerian subgraph. Let G_1 be a longest closed trail in G. Then G_1 must be G. If not, let $G_2 = G \backslash E(G_1)$. Since G is an edge-disjoint union of cycles, every vertex of G is of even degree ≥ 2. Further, since G_1 is Eulerian, each vertex of G_1 is of even degree ≥ 2. Hence, each vertex of G_2 is of even degree. Since G_2 is not totally disconnected and G is connected, G_2 contains a cycle C having a vertex v in common with G_1. Describe the Euler tour of G_1 starting and ending at v and follow it by C. Then $G_1 \cup C$ is a closed trail in G longer than G_1. This contradicts the choice of G_1 and so G_1 must be G. Hence, G is Eulerian. $\square$

If $G_1, \ldots, G_r$ are subgraphs of a graph G that are pairwise edge-disjoint and their union is G, then this fact is denoted by writing $G = G_1 \oplus \cdots \oplus G_r$. If in the above equation $G_i = C_i$, a cycle of G for each i, then $G = C_1 \oplus \cdots \oplus C_r$. The set of cycles $S = \{C_1, \ldots, C_r\}$ is then called a *cycle decomposition* of G. Thus, Theorem 6.1.2 implies that *a connected graph is Eulerian if, and only if, it admits a cycle decomposition.*

There is yet another characterization of Eulerian graphs due to McKee [87] and Toida [112]. Our proof is based on Fleischner [40], [41].

Theorem 6.1.3* *A graph is Eulerian if, and only if, each edge e of G belongs to an odd number of cycles of G.*

For instance, in Figure 6.3, e belongs to the three cycles $P_1 \cup e$, $P_2 \cup e$, and $P_3 \cup e$.

Proof Denote by γ_e the number of cycles of G containing e. Assume that γ_e is odd for each edge e of G. Since a loop at any vertex v of G is in exactly one cycle of G and contributes 2 to the degree of v in G, we may suppose that G is loopless.

Let $S = \{C_1, \ldots, C_m\}$ be the set of cycles of G. Replace each edge e of G by γ_e parallel edges, and replace e in each of the γ_e cycles containing e by a parallel edge, making sure that none of the parallel edges is repeated. Let the resulting graph

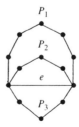

FIGURE 6.3. Eulerian graph with edge e belonging to three cycles

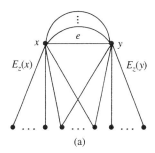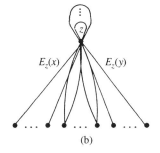

(a)   (b)

FIGURE 6.4. Graph for proof of Theorem 6.1.3

be G_o, and let the new set of cycles be $S_o = \{C_1^0, \ldots, C_m^o\}$. Clearly, S_o is a cycle decomposition of G_o. Hence by Theorem 6.1.2, G_o is Eulerian. But then $d_{G_o}(v) \equiv 0$ (mod 2) for each $v \in V(G_o) = V(G)$. Moreover, $d_G(v) = d_{G_o}(v) - \Sigma_e(\gamma_e - 1)$, where e is incident at v in G and hence $d_G(v) \equiv 0$ (mod 2), γ_e being odd for each $e \in E(G)$. Thus, G is Eulerian.

Conversely, assume that G is Eulerian. We proceed by induction on $n = |V(G)|$. If $n = 1$, each edge is a loop and hence belongs to exactly one cycle of G.

Assume the result for graphs with less than n (≥ 2) vertices. Let G be a graph with n vertices. Let $e = xy$ be an edge of G, and let $\lambda(e)$ be the multiplicity of e in G.

The graph $G.e$ obtained from G by contracting the edge e (cf. Section 4.3 of Chapter IV) is also Eulerian. Denote by z the new vertex of $G.e$ obtained by identifying the vertices x and y of G. The set of edges incident with z in $G.e$ is partitioned into three subsets (see Figure 6.4):

1. $E_z(x) =$ set of edges of G incident with x but not with y,
2. $E_z(y) =$ set of edges of G incident with y but not with x, and
3. $E_z(xy) =$ set of $\lambda(e) - 1$ loops of $G.e$ corresponding to the edges parallel to e in G.

Let $k = |E_z(x)|$. Then, since G is Eulerian,

$$k + \lambda(e) = d_G(x) \equiv 0 \pmod 2. \tag{1}$$

Let Γ_f and $\Gamma(e_i, e_j)$ denote, respectively, the number of cycles in $G.e$ containing the edge f and the pair (e_i, e_j) of edges. Since $|V(G.e)| = n - 1$, by the induction assumption, Γ_f is odd for each edge f of $G.e$. Now, any cycle of G containing e either consists of e and an edge parallel to e in G (and there are $\lambda(e) - 1$ of them) or it contains e, an edge e_i of $E_z(x)$ and an edge e'_j of $E_z(y)$. These correspond in $G.e$, respectively, to a loop at z and to a cycle containing the edges of $G.e$ that correspond to the edges e_i and e'_j of G. By abuse of notation, we denote these corresponding edges of $G.e$ also by e_i and e'_j, respectively. Moreover, any cycle of $G.e$ containing an edge e_i of $E_z(x)$ will also contain either an edge e_j of $E_z(x)$ or an edge e'_j of $E_z(y)$, but not both. A cycle of the former type is counted once in

Γ_{e_i} and once in Γ_{e_j} and these will not give rise to cycles in G containing e. Thus

$$\gamma_e = (\lambda(e) - 1) + \sum_{e_i \in E_z(x)} \Gamma_{e_i} - \sum_{\substack{i \neq j \\ e_i, e_j \in E_z(x)}} \Gamma(e_i, e_j).$$

Now, by induction hypothesis, $\Gamma_{e_i} \equiv 1 \pmod{2}$ for each e_i, and $\Gamma(e_i, e_j) = \Gamma(e_j, e_i)$ in the last sum on the right. Thus $\gamma_e \equiv (\lambda(e) - 1) + k \pmod{2}$, which is $\equiv 1 \pmod{2}$ by relation (1). □

A consequence of Theorem 6.1.3 is a result of Bondy and Halberstam [20], which gives yet another characterization of Eulerian graphs.

Corollary 6.1.4*   *A graph is Eulerian if, and only if, it has an odd number of cycle decompositions.*

Proof   In one direction, the proof is trivial. If G has an odd number of cycle decompositions, then it has at least one, and hence G is Eulerian.

Conversely, assume that G is Eulerian. Let $e \in E(G)$, and let $C_1, \ldots, C_r$ be the cycles containing e. By Theorem 6.1.3, r is odd. We proceed by induction on $m = |E(G)|$, with G Eulerian.

If G is just a cycle, then the result is true. Assume, then, that G is not a cycle. This means that for each i, $1 \leq i \leq r$, by the induction assumption, $G_i = G - E(C_i)$ has an odd number, say s_i, of cycle decompositions. (If G_i is disconnected, apply the induction assumption to each of the nontrivial components of G_i.) The union of each of these cycle decompositions of G_i and C_i yields a cycle decomposition of G. Hence, the number of cycle decompositions of G containing C_i is s_i, $1 \leq i \leq r$. Let $s(G)$ denote the number of cycle decompositions of G. Then

$$s(G) \equiv \sum_{i=1}^{r} s_i \equiv r \pmod{2} \quad (\text{since } s_i \equiv 1 \pmod{2})$$

$$\equiv 1 \pmod{2}. □$$

Exercise 1.1   Find an Euler tour in the graph G below.

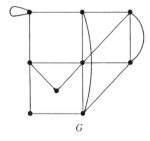

G

Exercise 1.2   Does there exist an Eulerian graph with
(i)   an even number of vertices and an odd number of edges?

(ii) an odd number of vertices and an even number of edges? Draw such a graph if it exists.

Exercise 1.3 Prove that a connected graph is Eulerian if, and only if, each of its blocks is Eulerian.

Exercise 1.4 If G is a connected graph with $2k(k > 0)$ vertices of odd degree, show that $E(G)$ can be partitioned into k open (i.e., not closed) trails.

Exercise 1.5 Prove that a connected graph is Eulerian if, and only if, each of its edge cuts has an even number of edges.

6.2. Hamiltonian Graphs

Definition 6.2.1 A graph is called *Hamiltonian* if it has a spanning cycle (see Figure 6.5(a)). These graphs were first studied by Sir William Hamilton, a mathematician. A spanning cycle of a graph G, when it exists, is often called a *Hamilton cycle* (or a *Hamiltonian cycle*) of G.

Definition 6.2.2 A graph G is called *traceable* if it has a spanning path of G (See Figure 6.5(b)). A spanning path of G is also called a *Hamilton path* (or a *Hamiltonian path*) of G.

Hamilton's "Around the World" Game

Hamilton introduced these graphs in 1859 through a game that used a solid dodecahedron (Figure 6.6). A dodecahedron has 20 vertices and 12 pentagonal faces. At each vertex of the solid, a peg was attached. The vertices were marked Amsterdam, Ann Arbor, Berlin, Budapest, Dublin, Edinburgh, Jerusalem, London, Melbourne, Moscow, Novosibirsk, New York, Paris, Peking, Prague, Rio di Janeiro, Rome, San Francisco, Tokyo, and Warsaw. Further, a string was also provided. The object of the game was to start from any one of the vertices and keep on attaching the string to the pegs as we move from one vertex to another along a particular edge with the condition that we have to get back to the starting city without visiting any intermediate city more than once. In other words, the problem asks one to find a Hamilton cycle in the graph of the dodecahedron (see Figure 6.6). Hamilton solved this problem as follows: When a traveler arrives at a city, he has the choice of taking

(a)

(b)

FIGURE 6.5. (a) Hamiltonian graph; (b) non-Hamiltonian but traceable graph

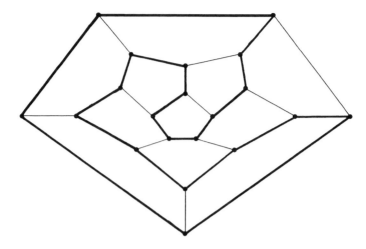

FIGURE 6.6. Solid dodecahedron for Hamilton's "around the world" problem

the edge to his right or left. Denote the choice of taking the edge to the right by R and the edge to the left by L. Let 1 denote the operation of staying where he is.

Define the product $O_1 O_2$ of two operations O_1 and O_2 as O_1 followed by O_2. For example, LR denotes going first left and then going right. Two sequences of operations are *equal* if, after starting at a vertex, the two sequences lead to the same vertex. The product defined above is associative but not commutative. Further, it is clear (see Figure 6.6) that

$$R^5 = L^5 = 1,$$

$$RL^2R = LRL,$$

$$LR^2L = RLR,$$

$$RL^3R = L^2, \text{ and}$$

$$LR^3L = R^2.$$

These relations give

$$1 = R^5 = R^2R^3 = (LR^3L)R^3 = (LR^3)(LR^3) = (LR^3)^2 = (LR^2R)^2$$
$$= (L(LR^3L)R)^2 = (L^2R^3LR)^2 = (L^2((LR^3L)R)LR)^2 = (L^3R^3LRLR)^2$$
$$= LLLRRRLRLRLLLRRRLRLR. \qquad (*)$$

The last sequence of operations contains 20 operations and contains no partial sequence equal to 1. Hence, this sequence must represent a Hamilton cycle. Thus starting from any vertex and following the sequence of operations $(*)$, we do indeed get a Hamilton cycle of the graph of Figure 6.6.

Knight's Tour in a Chess Board 6.2.3   The knight's tour problem is the problem of determining a closed tour through all the 64 squares of an 8×8 chessboard

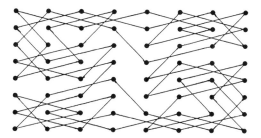

FIGURE 6.7. A knight's tour in a chessboard

by a knight with the condition that the knight does not visit any intermediate square more than once. This is equivalent to finding a Hamilton cycle in the corresponding graph of 64 ($= 8 \times 8$) vertices in which two vertices are adjacent if, and only if, the knight can move from one vertex to the other following the rules of the chess game. Figure 6.7 displays a knight's tour.

Even though Eulerian graphs admit an elegant characterization, no decent characterization of Hamiltonian graphs is known as yet. In fact, it is one of the most difficult unsolved problems in graph theory. (Actually, it is what is called an NP-complete problem; see reference [48].) Many sufficient conditions for a graph to be Hamiltonian are known; however, none of them happens to be an elegant necessary condition.

We begin with a necessary condition.

Theorem 6.2.4 *If G is Hamiltonian, then for every nonempty proper subset S of V, $\omega(G-S) \leq |S|$.*

Proof Let C be a Hamilton cycle in G. Then, since C is a spanning subgraph of G, $\omega(G-S) \leq \omega(C-S)$. If $|S| = 1$, $C-S$ is a path, and therefore $\omega(C-S) = 1 = |S|$. The removal of a vertex from a path P results in one or two components, according to whether the removed vertex is an end vertex or an internal vertex of P, respectively. Hence, by induction, the number of components in $C-S$ cannot exceed $|S|$. This proves that $\omega(G-S) \leq \omega(C-S) \leq |S|$. □

It follows directly from the definition of a Hamiltonian graph or from Theorem 6.2.4 that any Hamiltonian graph must be 2-connected. (If G has a cut vertex v, then taking $S = \{v\}$, we see that $\omega(G-S) > |S|$). The converse, however, is not true. For example, the theta graph of Figure 6.8 is 2-connected but not Hamiltonian. Here, P stands for a u–v path of any length ≥ 2 containing neither x nor y.

Exercise 2.1 Show by means of an example that the condition in Theorem 6.2.4 is not sufficient for G to be Hamiltonian.

Exercise 2.2 Use Theorem 6.2.4 to show that the Herschel graph (shown in Figure 5.4) is non-Hamiltonian.

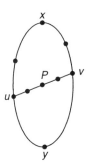

FIGURE 6.8. Theta graph

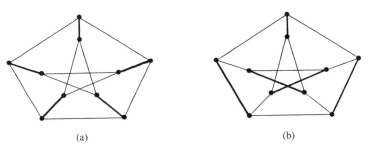

(a)   (b)

FIGURE 6.9. Petersen graph. The solid edges form a 1-factor of P

Exercise 2.3 Do Exercise 2.2 by using Theorem 1.4.10 (characterization theorem for bipartite graphs).

If a cubic graph G has a Hamilton cycle C, then $G\backslash E(C)$ is a 1-factor of G. Hence, for a cubic graph G to be Hamiltonian, G must have a 1-factor F such that $G\backslash E(F)$ is a Hamilton cycle of G. Now, the Petersen graph P (shown in Figure 1.7) has two different types of 1-factors (Figure 6.9), and for any such 1-factor F of P, $P\backslash E(F)$ consists of two disjoint 5-cycles. Hence P is non-Hamiltonian.

Theorem 6.2.5 is a basic result due to Ore [98], which gives a sufficient condition for a graph to be Hamiltonian.

Theorem 6.2.5 (Ore [98]) *Let G be a simple graph with $n \geq 3$ vertices. If for every pair of nonadjacent vertices u, v of G, $d(u) + d(v) \geq n$, then G is Hamiltonian.*

Proof Suppose that G satisfies the condition of the theorem but G is not Hamiltonian. Add edges to G (without adding vertices) and get a supergraph G^* of G such that G^* is a maximal simple graph that satisfies the condition of the theorem but G^* is non-Hamiltonian. Such a graph G^* must exist since G is non-Hamiltonian while the complete graph on $V(G)$ is Hamiltonian. Hence, for any pair u and v of nonadjacent vertices of G^*, $G^* + uv$ must contain a Hamilton cycle

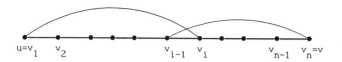

FIGURE 6.10. Hamilton path for proof of Theorem 6.2.5

C. This cycle C would certainly contain the edge $e = uv$. Then $C-e$ is a Hamilton path $u = v_1 v_2 v_3 \ldots v_n = v$ of G^* (see Figure 6.10).

Now, if $v_i \in N(u)$, $v_{i-1} \notin N(v)$; otherwise, $v_1 v_2 \cdots v_{i-1} v_n v_{n-1} v_{n-2} \cdots v_{i+1} v_i v_1$ would be a Hamilton cycle in G^*. Hence, for each vertex adjacent to u, there is a vertex of $V-\{v\}$ nonadjacent to v. But then

$$d_{G^*}(v) \leq (n-1) - d_{G^*}(u).$$

This gives that $d_{G^*}(u) + d_{G^*}(v) \leq n - 1$, and therefore $d_G(u) + d_G(v) \leq n - 1$, a contradiction.   □

Corollary 6.2.6 (Dirac [35])   *If G is a simple graph with $n \geq 3$ and $\delta \geq \frac{n}{2}$, then G is Hamiltonian.*   □

Corollary 6.2.7   *Let G be a simple graph with $n \geq 3$ vertices. If $d(u)+d(v) \geq n - 1$ for every pair of nonadjacent vertices u and v of G, then G is traceable.*

Proof   Choose a new vertex w and let G' be the graph $G \vee \{w\}$. Then each vertex of G has its degree increased by one, and therefore, in G', $d(u) + d(v) \geq n + 1$ for every pair of nonadjacent vertices. Since $|V(G')| = n + 1$, by Theorem 6.2.5, G' is Hamiltonian. If C' is a Hamilton cycle of G', then $C'-w$ is a Hamilton path of G. Thus, G is traceable.   □

Exercise 2.4   Show by means of an example that the conditions of Theorem 6.2.5 and its Corollary 6.2.6 are not necessary for a simple connected graph to be Hamiltonian.

Exercise 2.5   Show that if a cubic graph G has a spanning closed trial, then G is Hamiltonian.

Exercise 2.6   Prove that the n-cube Q_n is Hamiltonian for every $n \geq 2$.

Exercise 2.7   Prove that the wheel W_n is Hamiltonian for every $n \geq 4$.

Exercise 2.8   Prove that a simple k-regular graph on $2k - 1$ vertices is Hamiltonian.

Exercise 2.9   For any vertex v of the Petersen graph P, show that $P-v$ is Hamiltonian. (A non-Hamiltonian graph G with this property, namely, for any vertex v of G, $G-v$ is Hamiltonian, is called a hypo-Hamiltonian graph. In fact, P is the smallest graph with this property.)

Exercise 2.10 For any vertex v of P, show that there exists a Hamilton path starting at v.

Exercise 2.11 If $G = G(X, Y)$ is a bipartite Hamiltonian graph, then show that $|X| = |Y|$.

Exercise 2.12 Let G be a simple graph on $2k$ vertices with $\delta(G) \geq k$. Show that G has a perfect matching.

Exercise 2.13 Prove that a simple graph of order n with n even and $\delta \geq \frac{(n+2)}{2}$ has a 3-factor.

Bondy and Chvátal [18] observed that the proof of Theorem 6.2.5 is essentially based on the following result.

Theorem 6.2.8 *Let G be a simple graph of order $n \geq 3$ vertices. Then G is Hamiltonian if, and only if, $G + uv$ is Hamiltonian for every pair of nonadjacent vertices u and v with $d(u) + d(v) \geq n$.*

The last result has been instrumental for Bondy and Chvátal's definition of the closure of a graph G.

Definition 6.2.9 The *closure* of a graph G, denoted cl(G) is defined to be that supergraph of G obtained from G by recursively joining pairs of nonadjacent vertices whose degree sum is at least n until no such pair exists.

This recursive definition does not stipulate the order in which the new edges are added. Hence, we must show that the definition does not depend upon the order of the newly added edges. Figure 6.11 explains the construction of cl(G).

Theorem 6.2.10 *The closure cl(G) of a graph G is well-defined.*

Proof Let G_1 and G_2 be two graphs obtained from G by recursively joining pairs of nonadjacent vertices whose degree sum is at least n until no such pair exists. We have to prove that $G_1 = G_2$.

Let $\{e_1, \ldots, e_p\}$ and $\{f_1, \ldots, f_q\}$ be the sets of new edges added to G to get G_1 and G_2, respectively. We want to show that each e_i is some f_j (and therefore belongs to G_2) and that each f_k is some e_ℓ (and therefore belongs to G_1). Let e_i be the first edge in $\{e_1, \ldots, e_p\}$ not belonging to G_2. Then $e_1, \ldots, e_{i-1}$ are all in both G_1 and G_2, and $uv = e_i \notin E(G_2)$. Let $H = G + \{e_1, \ldots, e_{i-1}\}$. Then H is a subgraph of both G_1 and G_2. By the way, cl(G) is defined,

$$d_H(u) + d_H(v) \geq n,$$

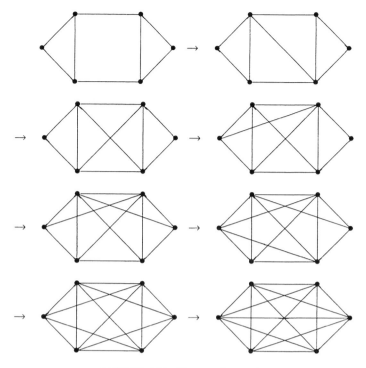

FIGURE 6.11. Closure of a graph

and hence,

$$d_{G_2}(u) + d_{G_2}(v) \geq n.$$

But this is a contradiction since u and v are nonadjacent vertices of G_2, and G_2 is a closure of G. Thus, $e_i \in E(G_2)$ and, similarly, each $f_k \in E(G_1)$. $\square$

An immediate consequence of Theorem 6.2.8 is the following.

Theorem 6.2.11   *If* cl(G) *is Hamiltonian, then G is Hamiltonian.*

Corollary 6.2.12   *If* cl(G) *is complete, then G is Hamiltonian.*

Exercise 2.14 Determine the closure of the following graph.

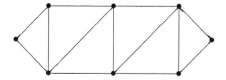

We conclude this section with a result of Chvátal and Erdös [28].

Theorem 6.2.13 (Chvátal and Erdös)   *If for a simple 2-connected graph G,*
$\alpha \leq \kappa$, *then G is Hamiltonian. (α is the independence number of G and κ is the*
connectivity of G.)

Proof   Suppose $\alpha \leq \kappa$ but G is not Hamiltonian. Let $C : v_0 v_1 \ldots v_{p-1}$ be a
longest cycle of G. We fix this orientation on C. By Dirac's theorem (Exercise
5.4 of Chapter III), $p \geq \kappa$. Let $v \in V(G) \backslash V(C)$. Then by Menger's theorem (see
also Exercise 5.3 of Chapter III) there exist κ internally disjoint paths $P_1, \ldots, P_\kappa$
from v to C. Let $v_{i_1}, v_{i_2}, \ldots, v_{i_\kappa}$ be the end vertices (with suffixes in the increasing
order) of these paths on C. No two of the consecutive vertices $v_{i_1}, v_{i_2}, \ldots, v_{i_\kappa}, v_{i_1}$
can be adjacent vertices of C, since otherwise we get a cycle of G longer than C.
Hence, between any two consecutive vertices of $\{v_{i_1}, v_{i_2}, \ldots, v_{i_\kappa}, v_{i_1}\}$, there exists
at least one vertex of G. Let u_{i_j} be the vertex next to v_{i_j} in the $v_{i_j} - v_{i_{j+1}}$ path along
C (see Figure 6.12(a)).
We claim that $\{u_{i_1}, \ldots, u_{i_\kappa}\}$ is an independent set of G. Suppose u_{i_j} is adjacent
to $u_{i_m}, m > j$ (suffixes taken modulo k); then

$$u_{i_j} \cdots v_{i_{j+1}} \cdots v_{i_m} v v_{i_j} \cdots v_{i_{j-1}} \cdots u_{i_m} u_{i_j}$$

is a cycle of G longer than C, a contradiction.

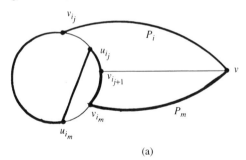

(a)

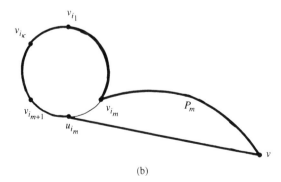

(b)

FIGURE 6.12. Graphs for proof of Theorem 6.2.13

Clearly, $\{v, u_{i_1}, \ldots, u_{i_\kappa}\}$ is an independent set of G. (Otherwise, $vu_{i_m} \in E(G)$ for some m. See Figure 6.12(b). Then

$$vu_{i_m} \cdots v_{i_{m+1}} \cdots v_{i_\kappa} \cdots v_{i_1} \cdots v_{i_m} P_m^{-1} v$$

is a cycle longer than C, a contradiction.) But then $\alpha > \kappa$, a contradiction to our hypothesis. Thus G is Hamiltonian. $\square$

This theorem, although interesting, is not powerful in that for the cycle C_n, $\kappa = 2$ while $\alpha = \lfloor \frac{n}{2} \rfloor$ and hence increases with n.

A graph G with at least three vertices is *Hamiltonian-connected* if any two vertices of G are connected by a Hamilton path in G. For example, for $n \geq 3$, K_n is Hamiltonian-connected, whereas for $n \geq 4$, C_n is not Hamiltonian-connected.

Theorem 6.2.14 *If G is a simple graph with $n \geq 3$ vertices such that $d(u) + d(v) \geq n+1$ for every pair of nonadjacent vertices of G, then G is Hamiltonian-connected.*

Proof Let u and v be any two vertices of G. Our aim is to show that there exists a Hamilton path from u to v in G.

Choose a new vertex w, and let $G^* = G \cup \{wu, wv\}$. We claim that cl $(G^*) = K_{n+1}$. First, the recursive addition of the pairs of nonadjacent vertices u and v of G with $d(u) + d(v) \geq n + 1$ gives K_n. Further, each vertex of K_n is of degree $n - 1$ in K_n and $d_{G^*}(w) = 2$. Hence, cl $(G^*) = K_{n+1}$. So, by Corollary 6.2.12, G^* is Hamiltonian. Let C be a Hamilton cycle in G^*. Then $C - w$ is a Hamilton path in G from u to v. $\square$

6.3.* Pancyclic Graphs

Definition 6.3.1 A graph G of order $n(\geq 3)$ is *pancyclic* if G contains cycles of all lengths from 3 to n. G is called *vertex-pancyclic* if each vertex v of G belongs to a cycle of every length ℓ, $3 \leq \ell \leq n$.

Example 6.3.2 Clearly, *a vertex-pancyclic graph is pancyclic*. However, the converse is not true. Figure 6.13 displays a pancyclic graph that is not vertex-pancyclic.

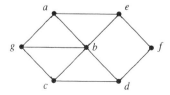

FIGURE 6.13. Pancyclic graph that is not vertex-pancyclic

The study of pancyclic graphs was initiated by Bondy [18], who showed that Ore's sufficient condition for a graph G to be Hamiltonian (Theorem 6.2.5) actually implies much more. Note that if $\delta \geq \frac{n}{2}$, then $m \geq \frac{n^2}{4}$.

Theorem 6.3.3 *Let G be a simple Hamiltonian graph on n vertices with at least $\lfloor \frac{n^2}{2} \rfloor$ edges. Then G is either pancyclic or else is the complete bipartite graph $K_{\frac{n}{2},\frac{n}{2}}$. In particular, if G is Hamiltonian and $m > \frac{n^2}{4}$, then G is pancyclic.*

Proof The result can directly be verified for $n = 3$. We may therefore assume that $n \geq 4$.

We apply induction on n. Suppose the result is true for all graphs of order at most $n - 1 (n \geq 4)$, and let G be a graph of order n.

First, assume that G has a cycle $C = v_0 v_1 \ldots v_{n-2} v_0$ of length $n - 1$. Let v be the (unique) vertex of G not belonging to C. If $d(v) \geq \frac{n}{2}$, v is adjacent to two consecutive vertices on C, and hence G has a cycle of length 3. Suppose for some $r, 2 \leq r \leq \frac{n}{2}$, C has no pair of vertices u and w on C adjacent to v in G with $d_C(u, w) = r$. Then, if $v_{i_1}, v_{i_2}, \ldots v_{i_{d(v)}}$ are the vertices of C that are adjacent to v in G (recall that C contains all the vertices of G except v), then $v_{i_1+r}, v_{i_2+r}, \ldots, v_{i_{d(v)}+r}$ are nonadjacent to v in G, where the suffixes are taken modulo $(n - 1)$. Hence, $2d(v) \leq n - 1$, a contradiction. Hence, for each $r, 2 \leq r \leq \frac{n-1}{2}$, C has a pair of vertices u and w on C adjacent to v in G with $d_C(u, w) = r$. Thus, for each $r, 2 \leq r \leq \frac{n-1}{2}$, G has a cycle of length $r + 2$ as well as a cycle of length $n - 1 - r + 2 = n - r + 1$. (See Figure 6.14.) Thus G is pancyclic.

If $d(v) \leq \frac{n-1}{2}$, then $G[V(C)]$, the subgraph of G induced by $V(C)$ has at least $\frac{n^2}{4} - \frac{n-1}{2} > \frac{(n-1)^2}{4}$ edges. So, by the induction assumption, $G[V(C)]$ is pancyclic and hence G is pancyclic. (By hypothesis, G is Hamiltonian.)

Next, assume that G has no cycle of length $n - 1$. Then G is not pancyclic. In this case, we show that G is $K_{\frac{n}{2},\frac{n}{2}}$.

Let $C = v_0 v_1 v_2 \cdots v_{n-1} v_0$ be a Hamilton cycle of G. We claim that of the two pairs $v_i v_k$ and $v_{i+1} v_{k+2}$ (where suffixes are taken modulo n), at most only one of them can be an edge of G. Otherwise, $v_k v_{k-1} v_{k-2} \cdots v_{i+1} v_{k+2} v_{k+3} v_{k+4} \cdots v_i v_k$ is an $(n-1)$-cycle in G, a contradiction. Hence, if $d(v_i) = r$, then there are r vertices adjacent to v_i in G and hence at least r vertices (including v_{i+1} since $v_i v_{i-1} \in E(G)$) that are nonadjacent to v_{i+1}. Thus, $d(v_{i+1}) \leq n - r$, and $d(v_i) + d(v_{i+1}) \leq n$.

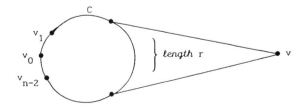

FIGURE 6.14. Graph for proof of Theorem 6.3.3

Summing the last inequality over i from 0 to $n - 1$, we get $4m \leq n^2$. But, by hypothesis, $4m \geq n^2$. Hence, $m = \frac{n^2}{4}$ and so n must be even. Again, this yields $d(v_i) + d(v_{i+1}) = n$ for each i, and thus, for each i and k,

$$\text{exactly one of } v_i v_k \text{ and } v_{i+1} v_{k+2} \text{ is an edge of } G. \tag{$*$}$$

Thus, if $G \neq K_{\frac{n}{2}, \frac{n}{2}}$, then certainly there exist i and j such that $v_i v_j \in E$ and $i \equiv j \pmod 2$. Hence, for some j, there exists an even positive integer s such that $v_{j+1} v_{j+1+s} \in E$. Choose s to be the least even positive integer with the above property. Then $v_j v_{j+s-1} \notin E$. Hence $s \geq 4$ (as $s = 2$ would mean that $v_j v_{j+1} \notin E$). Again, by $(*)$, $v_{j-1} v_{j+s-3} = v_{j-1} v_{j-1+(s-2)} \in E(G)$ contradicting the choice of s. Thus $G = K_{\frac{n}{2}, \frac{n}{2}}$. The last part follows from the fact that $|E(K_{\frac{n}{2}, \frac{n}{2}})| = \frac{n^2}{4}$. $\quad\square$

Corollary 6.3.4 *Let $G \neq K_{\frac{n}{2}, \frac{n}{2}}$, be a simple graph with $n \geq 3$ vertices, and let $d(u) + d(v) \geq n$ for every pair of nonadjacent vertices of G. Then G is pancyclic.*

Proof By Ore's Theorem (Theorem 6.2.5), G is Hamiltonian. We show that G is pancyclic by first proving that $m \geq \frac{n^2}{4}$ and then invoking Theorem 6.3.3. This is true if $\delta \geq \frac{n}{2}$ (as $2m = \sum_{i=1}^{n} d_i \geq \delta n \geq n^2/2$). So assume that $\delta < \frac{n}{2}$.

Let S be the set of vertices of degree δ in G. For every pair (u, v) of vertices of degree δ, $d(u) + d(v) < \frac{n}{2} + \frac{n}{2} = n$. Hence by hypothesis, S induces a clique of G and $|S| \leq \delta + 1$. If $|S| = \delta + 1$, then G is disconnected with $G[S]$ as a component, which is impossible (as G is Hamiltonian). Thus $|S| \leq \delta$. Further, if $v \in S$, v is nonadjacent to $n - 1 - \delta$ vertices of G. If u is such a vertex, $d(v) + d(u) \geq n$ implies that $d(u) \geq n - \delta$. Further v is adjacent to at least one vertex $w \notin S$ and $d(w) \geq \delta + 1$, by the choice of S. These facts give that

$$2m = \sum_{i=1}^{n} d_i \geq (n - \delta - 1)(n - \delta) + \delta^2 + (\delta + 1),$$

where the last $(\delta + 1)$ comes out of the degree of w. Thus

$$2m \geq n^2 - n(2\delta + 1) + 2\delta^2 + 2\delta + 1$$

which implies that

$$4m \geq 2n^2 - 2n(2\delta + 1) + 4\delta^2 + 4\delta + 2$$
$$= (n - (2\delta + 1))^2 + n^2 + 1$$
$$\geq n^2 + 1, \quad \text{since } n > 2\delta.$$

Consequently, $m > \frac{n^2}{4}$, and by Theorem 6.3.3, G is pancyclic. $\quad\square$

6.4.   Hamilton Cycles in Line Graphs

We now turn our attention to the existence of Hamilton cycles in line graphs.

Theorem 6.4.1   *If G is Eulerian, then $L(G)$, the line graph of G, is both Hamiltonian and Eulerian.*

Proof   If $e_1 e_2 \cdots e_m$ is the edge sequence of an Euler tour in G, and if vertex u_i in $L(G)$ represents the edge e_i, $1 \leq i \leq m$, then $u_1 u_2 \cdots u_m u_1$ is a Hamilton cycle of $L(G)$. Further, if $e = v_1 v_2 \in E(G)$ and the vertex u in $L(G)$ represents the edge e, then $d_{L(G)}(u) = d_G(v_1) + d_G(v_2) - 2$, which is even (and ≥ 2) since both $d_G(v_1)$ and $d_G(v_2)$ are even (and ≥ 2). Hence, in $L(G)$ every vertex is of even degree (≥ 2). So $L(G)$ is also Eulerian.   □

Exercise 4.1   Disprove the converse of Theorem 6.4.1 by a counterexample.

Definition 6.4.2   A *dominating trail* of a graph G is a closed trail T in G (which may be just a single vertex) such that every edge of G not in T is incident with T.

Example 6.4.3   For instance, in the graph of Figure 6.13, the trail *abcdbea* is a dominating trail.

Harary and Nash-Williams [63] characterized graphs that have Hamiltonian line graphs.

Theorem 6.4.4 (Harary–Nash-Williams)   *The line graph of a graph G with at least three edges is Hamiltonian if, and only if, G has a dominating trail.*

Proof   Let T be a dominating trail of G, and let $\{e_1, e_2, \ldots, e_s\}$ be the edge sequence representing T. Assume that e_1 and e_2 are incident at v_1. Replace the subsequence $\{e_1, e_2\}$ of $\{e_1, e_2, \ldots, e_s\}$ by the sequence $\{e_1, e_{11}, e_{12}, \ldots, e_{1r_1}, e_2\}$, where $e_{11}, e_{12}, \ldots, e_{1r_1}$ are the edges incident at v_1 other than e_1 and e_2. Assume that we have already replaced the subsequence $\{e_i, e_{i+1}\}$ by $\{e_i, e_{i1}, \ldots, e_{ir_i}, e_{i+1}\}$. Then replace $\{e_{i+1}, e_{i+2}\}$ by the sequence $\{e_{i+1}, e_{(i+1)1}, \ldots, e_{(i+1)r_{(i+1)}}, e_{i+2}\}$, where the new edges $e_{(i+1)1}, \ldots, e_{(i+1)r_{(i+1)}}$ have not appeared in the previous i subsequences. (Here we take $e_{s+1} = e_1$.) The resulting edge sequence is $e_1 e_{11} e_{12} \ldots e_{1r_1}$ $e_2 e_{21} e_{22} \ldots e_{2r_2} e_3 \ldots e_s e_{s1} e_{s2} \ldots e_{sr_s} e_1$ and this gives the Hamilton cycle $u_1 u_{11} u_{12}$ $\ldots u_{1r_1} u_2 u_{21} u_{22} \ldots u_{2r_2} u_3 \ldots u_s u_{s1} u_{s2} \ldots u_{sr_s} u_1$ in $L(G)$. (Here, u_1 is the vertex of $L(G)$ that corresponds to the edge e_1 of G, and so on)

Conversely, assume that $L(G)$ has a Hamilton cycle C. Let $C = u_1 u_2 \ldots u_m u_1$ and let e_i be the edge of G corresponding to the vertex u_i of $L(G)$. Let T_0 be the edge sequence $e_1 e_2 \ldots e_m e_1$. We now delete edges from T_0 one after another as follows: Let $e_i e_j e_k$ be the first three distinct consecutive edges of T_0 that have a common vertex; then delete e_j from the sequence. Let $T_0' = T_0 - e_j = e_1 e_2 \ldots e_i e_k \ldots e_m e_1$.

Now proceed with T_0' as we did with T_0. Continue this process until no such triad of edges exists. Then the resulting subsequence of T_0 must be a dominating trail or a pair of adjacent edges incident at a vertex, say, v_0. In the latter case, all the edges of G are incident at v_0, and hence we take v_0 as the dominating trail of G. □

Corollary 6.4.5 *The line graph of a Hamiltonian graph is Hamiltonian.*

Proof Let G be a Hamiltonian graph with Hamilton cycle C. Then C is a dominating trail of G. Hence $L(G)$ is Hamiltonian. □

Exercise 4.2 Show that the line graph of a graph G has a Hamilton path if, and only if, G has a trail T such that every edge of G not in T is incident with T. .

Exercise 4.3 Draw the line graph of the graph of Figure 6.13 and display a Hamilton cycle in it.

Theorem 6.4.6 (Balakrishnan and Paulraja [8]) *Let G be any connected graph. If each edge of G belongs to a triangle in G, then G has a spanning, Eulerian subgraph.*

Proof Since G has a triangle, G certainly has a closed trail. Let T be a longest closed trail in G. Then T must be a spanning Eulerian subgraph of G. If not, there exists a vertex v of G with $v \notin T$ and v is adjacent to a vertex u of T.

By hypothesis, uv belongs to a triangle, say uvw. If none of the edges of this triangle is in T then $T \cup \{uv, vw, wu\}$ yields a closed trail longer than T (See Figure 6.15). If $uw \in T$, then $(T-uw) \cup \{uv, vw\}$ would be a closed trail longer than T. This contradiction proves that T is a spanning closed trail of G. □

Corollary 6.4.7 *Let G be any connected graph. If each edge of G belongs to a triangle, then $L(G)$ is Hamiltonian.*

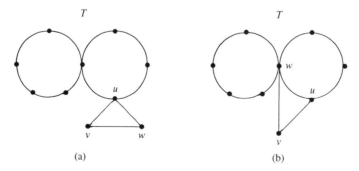

FIGURE 6.15. Graph for proof of Theorem 6.4.6. (a) $T \cup \{uv, vw, wu\}$ is longer than T; (b) $(T \backslash \{uw\}) \cup \{uv, vw\}$ is longer than T

Proof The proof is an immediate consequence of Theorems 6.4.4 and 6.4.6. □

Corollary 6.4.8 (Chartrand and Wall [24])   *If G is connected and $\delta(G) \geq 3$, then $L^2(G)$ is Hamiltonian.*

(Note: For $n > 1$, $L^n(G) = L(L^{n-1}(G))$, and $L^0(G) = G$.)

Proof   Since $\delta(G) \geq 3$, each vertex of $L(G)$ belongs to a clique of size at least three, and hence each edge of $L(G)$ belongs to a triangle. Now apply Corollary 6.4.7. □

Corollary 6.4.9 (Nebesky [95])   *If G is a connected graph with at least three vertices, then $L(G^2)$ is Hamiltonian.*

Proof   Since G is a connected graph with at least three vertices, every edge of G^2 belongs to a triangle. Hence, $L(G^2)$ is Hamiltonian by Corollary 6.4.7. □

Theorem 6.4.10   *Let G be a connected graph in which every edge belongs to a triangle. If e_1 and e_2 are edges of G such that $G \backslash \{e_1, e_2\}$ is connected, then there exists a spanning trail of G with e_1 and e_2 as its initial and terminal edges.*

Proof   The proof is essentially the same as for Theorem 6.4.6 and is based on considering the longest trail T in G with e_1 and e_2 as its initial and terminal edges, respectively. □

Corollary 6.4.11 (Balakrishnan and Paulraja [8])   *Let G be any connected graph with $\delta(G) \geq 4$. Then $L^2(G)$ is Hamiltonian-connected.*

Proof   The edges incident to a vertex v of G will yield a clique of size $d(v)$ in $L(G)$. Since $\delta(G) \geq 4$, $L(G)$ is 3-edge connected, and therefore for any pair of distinct edges e_1 and e_2 of $L(G)$, $L(G) \backslash \{e_1, e_2\}$ is connected. Further, each edge of $L(G)$ belongs to a triangle. Hence, there exists, by Theorem 6.4.10, a spanning trail T in $L(G)$ having e_1 and e_2 as initial and terminal edges. Hence, if there are any edges of $L(G)$ not belonging to T, they can only be "chords" of T. It follows (see Exercise 4.2) that in $L^2(G)$ there exists a Hamilton path starting and ending at the vertices corresponding to e_1 and e_2, respectively. Since e_1 and e_2 are arbitrary, $L^2(G)$ is Hamiltonian-connected. □

Corollary 6.4.12 (Jaeger [70])   *The line graph of a 4-edge-connected graph is Hamiltonian.*

To prove Corollary 6.4.12, we need the following lemma.

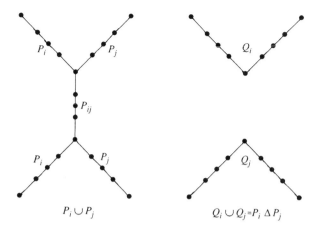

FIGURE 6.16. Graphs for proof of Lemma 6.4.13

Lemma 6.4.13 *Let S be a set of vertices of a nontrivial tree T, and let $|S| = 2k, k \geq 1$. Then there exists a set of k pairwise edge-disjoint paths whose end vertices are all the vertices of S.*

Proof Certainly, there exists a set of k paths in T whose end vertices are all the vertices of S. Choose such a set of k paths, say $\mathscr{P} = \{P_1, P_2, \ldots, P_k\}$ with the additional condition that the sum of their lengths is minimum.

We claim that the paths of $\mathscr{P}$ are pairwise edge-disjoint. If not, there exists a pair $\{P_i, P_j\}, i \neq j$, with P_i and P_j having an edge in common. In this case, P_i and P_j have path P_{ij} (of length ≥ 1) in common. Then $P_i \triangle P_j$, the symmetric difference of P_i and P_j, is a disjoint union of two paths, say Q_i and Q_j, with their end vertices being disjoint pairs of vertices belonging to S (Figure 6.16).

If we replace P_i and P_j by Q_i and Q_j in $\mathscr{P}$, then the resulting set of paths has the property that their end vertices are all the vertices of S and that the sum of their lengths is less than the sum of the lengths of the paths in $\mathscr{P}$. This contradicts the choice of $\mathscr{P}$, and hence $\mathscr{P}$ has the stated property. □

Proof of Corollary 6.4.12 Let G be a 4-edge connected graph. In view of Theorem 6.4.4, it suffices to show that G contains a spanning Eulerian subgraph.

By Theorem 4.3.4, G contains two edge-disjoint spanning trees T_1 and T_2. Let S be the set of vertices of odd degree in T_1. Then $|S|$ is even. Let $|S| = 2k, k \geq 1$. By Lemma 6.4.13, there exists a set of k pairwise edge-disjoint paths $\{P_1, P_2, \ldots, P_k\}$ in T_2 with the property stated in Lemma 6.4.13. Then $G_0 = T_1 \cup (P_1 \cup P_2 \cup \cdots \cup P_k)$ is a connected spanning subgraph of G in which each vertex is of even degree. Thus, G_0 is a spanning Eulerian subgraph of G. □

We conclude this section with a theorem on locally connected graphs.

Theorem 6.4.14*[Oberly and Sumner [97]] *A connected locally connected nontrivial $K_{1,3}$-free graph is Hamiltonian.*

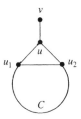

FIGURE 6.17. Graph for proof of Theorem 6.4.14

Proof Let G be a connected locally connected nontrivial $K_{1,3}$-free graph. We may assume that G has at least four vertices. Since G is locally connected, G certainly has a cycle. Let C be a longest cycle of G. If C is not a Hamilton cycle, there exists a vertex $v \in V(G)\setminus V(C)$ that is adjacent to a vertex u of C. Let u_1 and u_2 be the neighbors of u on C. Then, as the edges uv, uu_1, and uu_2 do not induce a $K_{1,3}$ in G, $u_1u_2 \in E(G)$, since otherwise v is adjacent either to u_1 or u_2 and we get a cycle longer than C, a contradiction. (See Figure 6.17.)

For each $x \in V(G)$, denote by $G_o(x)$ the subgraph $G[N_G(x)]$ of G. By hypothesis, $G_o(u)$ is connected, and hence there exists either a v–u_1 path P in $G_o(u)$ not containing u_2 or a v–u_2 path Q in $G_o(u)$ not containing u_1. Let us say it is the former. For the purposes of the proof of this theorem, we call a vertex $w \in V_o = (V(C) \cap V(P))\setminus\{u_1\}$ *singular* if none of the two neighbors of w on C is in $N_G(u)$.

Case 1. Each vertex of V_o is singular. Then for any $w \in V_o$, w is adjacent to u (since $w \in V(P) \subseteq V(G_o(u))$) but, since w is singular, none of the neighbors w_1 and w_2 of w on C is adjacent to u in G. Then, considering the $K_{1,3}$ subgraph $\{ww_1, ww_2, wu\}$, we see that $w_1w_2 \in E(G)$. Now, describe the cycle C' as follows: start from u_2, move away from u along C, and whenever we encounter a singular vertex w, bypass it by going through the edge w_1w_2. After reaching u_1, retrace the u_1–v path P^{-1} and follow it up by the path vuu_2. Then C' traverses through each vertex of $C \cup P$ exactly once. Thus, C' is a cycle longer than C, a contradiction to the choice of C. (See Figure 6.18).

Case 2. V_o has a nonsingular vertex. Let w be the first nonsingular vertex as P is traversed from v to u_1. As before, let w_1 and w_2 be the neighbors of w along C. Then at least one of w_1 and w_2 is adjacent to u. Without loss of generality, assume that w_2 is adjacent to u. Let

$$C' = (C \cup \{w_2u, uw, u_1u_2\})\setminus\{w_2w, uu_1, uu_2\}.$$

(See Figure 6.19.)

Clearly, C and C' are of the same length, and therefore C' is also a longest cycle of G. Then, by the choice w, the v–w section of P contains the only nonsingular vertex w. Let w_0 be the first singular vertex on this section. Consider the cycle C'' described as follows: start from w_2 and move along C' away from u until we reach the vertex preceding w_0. Bypass w_0 by moving through the neighbors of w_0 along C' (as in case 1), and repeat it for each nonsingular vertex after w_0. After

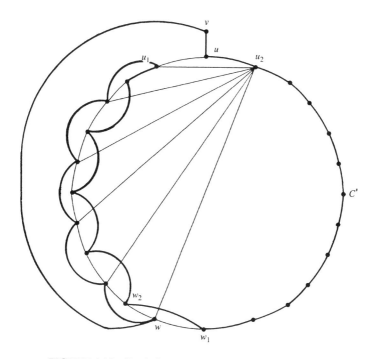

FIGURE 6.18. Graph for case 1 of proof of Theorem 6.4.14

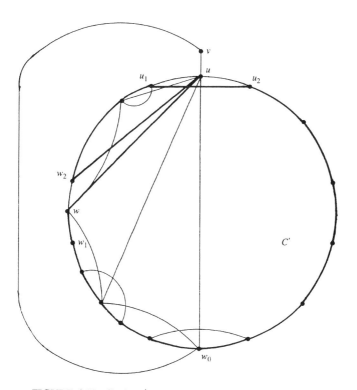

FIGURE 6.19. Cycle C' for case 2 of proof of Theorem 6.4.14

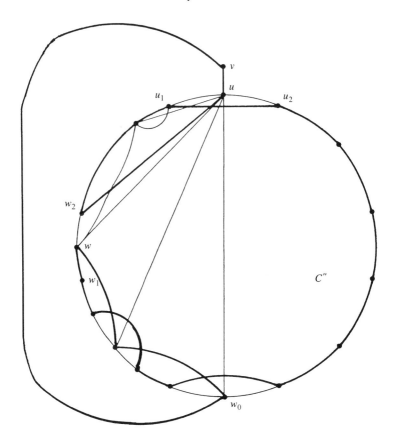

FIGURE 6.20. Cycle C'' for case 2 of proof of Theorem 6.4.14

reaching w, move along the w–v section of P^{-1} and follow it by the path vuw_2 (see Figure 6.20). Then C'' is a cycle longer than C' (as in case 1), a contradiction.

Hence, in any case, C cannot be a longest cycle of G. Thus, G is Hamiltonian.   □

6.5. 2-Factorable Graphs

It is clear that if a graph G is r-factorable with k r-factors, then the degree of each vertex of G is rk. In particular, if G is 2-factorable, then G is regular of even degree, say, $2k$. That the converse is also true is a result due to Petersen [104].

Theorem 6.5.1 (Petersen) *Every $2k$-regular graph, $k \geq 1$, is 2-factorable.*

Proof Let G be a $2k$-regular graph with $V = \{v_1, v_2, \ldots, v_n\}$. We may assume without loss of generality that G is connected. (Otherwise, we can consider the components of G separately.) Since each vertex of G is of even degree, by Theorem 6.1.2, G is Eulerian. Let T be an Euler tour of G. Form a bipartite graph H with bipartition (V, W), where $V = \{v_1, v_2, \ldots, v_n\}$ and $W = \{w_1, w_2, \ldots, w_n\}$ and in which v_i is made adjacent to w_j if, and only if, v_j follows v_i immediately in T. Since at every vertex of G there are k incoming edges and k outgoing edges along T, H is k-regular. Hence, by Theorem 5.4.3, H is 1-factorable. Let the k 1-factors be $M_1, \ldots, M_k$. Label the edges of M_i with the label i, $1 \le i \le k$. Then the k edges incident at each v_i receive the k labels $1, 2, \ldots, k$ and hence if the edges $v_i w_j$ and $v_j w_r$ are in M_p, identification of the vertex w_j with the vertex v_j for each j in M_p gives an edge labeling to G in which the edges $v_i v_j$ and $v_j v_r$ receive the label p. It is then clear that the edges of M_p yield a 2-factor of G with label p. Since this is true for each of the 1-factors M_p, $1 \le p \le k$, we get a 2-factorization of G into k 2-factors. $\quad\Box$

A special case of Theorem 6.5.1 is the 2-factorization of K_{2n+1}, which is $2n$-regular. Actually, K_{2n+1} has a 2-factorization into Hamilton cycles.

Theorem 6.5.2 K_{2n+1} *is 2-factorable into n Hamilton cycles.*

Proof Label the vertices of K_{2n+1} as $v_0, v_1, \ldots, v_{2n}$. For $i = 0, 1, \ldots, n$, let P_i be the path $v_i v_{i-1} v_{i+1} v_{i-2} v_{i+2} \cdots v_{i+n-1} v_{i-(n-1)}$ (suffixes taken modulo $2n$), and let C_i be the Hamilton cycle obtained from P_i by joining v_{2n} to the end vertices of P_i. The cycles C_i are edge-disjoint. This may be seen by placing the $2n$ vertices $v_0, v_1, \ldots, v_{2n-1}$ symmetrically on a circle and placing v_{2n} at the center of the circle and noting that the edges $v_i v_{i-1}, v_{i+1} v_{i-2}, \ldots, v_{i+n-1} v_{i-n+1}$ form a set of n parallel chords of this circle. $\quad\Box$

Figure 6.21 displays the three sets of parallel chords and three edge-disjoint Hamilton cycles in K_7.

The 2-factors are:
$$F_1 : v_6 v_0 v_5 v_1 v_4 v_2 v_3 v_6,$$
$$F_2 : v_6 v_1 v_0 v_2 v_5 v_3 v_4 v_6,$$
$$F_3 : v_6 v_2 v_1 v_3 v_0 v_4 v_5 v_6.$$

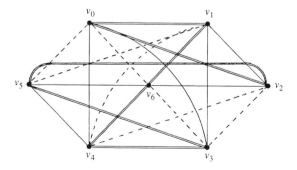

FIGURE 6.21. Parallel chords and edge-disjoint Hamilton cycles in K_7

6.6. Exercises

6.1   Prove: A Hamiltonian-connected graph is Hamiltonian. (Note: The converse is not true. See the next exercise.)

6.2   Show that a Hamiltonian-connected graph is 3-connected. Display a Hamiltonian graph of connectivity three that is not Hamiltonian-connected.

6.3   If G is traceable, show that for every proper subset S of $V(G)$, $\omega(G{-}S) \leq |S| + 1$. Disprove the converse by a counterexample.

6.4   If G is simple and $\delta \geq \frac{n-1}{2}$, show that G is traceable. Disprove the converse.

6.5   If G is simple and $\delta \geq \frac{n+1}{2}$, show that G is Hamiltonian-connected. Is the converse true?

6.6   Give an example of a non-Hamiltonian simple graph G of order $n(n \geq 3)$ such that for every pair of nonadjacent vertices u and v, $d(u) + d(v) \geq n - 1$. (This shows that the condition in Ore's Theorem (Theorem 6.2.5) cannot be weakened further.)

6.7   Show that if a cubic graph is Hamiltonian, then it has three disjoint 1-factors.

6.8*   Show that if a cubic graph has a 1-factor, then it has at least three distinct 1-factors.

6.9   Show that a complete k-partite graph G is Hamiltonian if, and only if, $|V(G)\backslash N| \geq |N|$, where N is the size of a maximum part of G. (See Aravamudhan and Rajendran [5].)

6.10   A graph is called *locally Hamiltonian* if $G[N(v)]$ is Hamiltonian for each vertex v of G. Show that a locally Hamiltonian graph is 3-connected.

6.11   Prove that $L(G)$ is locally Hamiltonian if, and only if, $G \cong K_{1,n}, n \geq 4$.

6.12   If G is a 2-connected graph that is both $K_{1,3}$-free and $(K_{1,3} + e)$-free, then prove that G is Hamiltonian. (Recall that a graph G is *H-free* if G does not contain an isomorphic copy of H as an induced subgraph.)

6.13   Let G be a simple graph of order $2n(n \geq 2)$. If for every pair of nonadjacent vertices u and v, $d(u) + d(v) \geq 2n + 2$, show that G contains a spanning cubic graph.

6.14   Show by means of an example that the square of a 1-connected (i.e., connected) graph need not be Hamiltonian. (A celebrated result of H. Fleischner states that the square of any 2-connected graph is Hamiltonian—a result that was originally conjectured by M. D. Plummer.)

6.15* Let G be a simple graph with degree sequence $(d_1, d_2, \ldots, d_n)$, where $d_1 \leq d_2 \leq \ldots \leq d_n$ and $n \geq 3$. Suppose that there is no value of r less than $\frac{n}{2}$ for which $d_r \leq r$ and $d_{n-r} < n - r$. Then show that G is Hamiltonian. (See Chvátal [26] or reference [19].)

6.16   Does there exist a simple non-Hamiltonian graph with degree sequence (2, 3, 5, 5, 5, 6, 6, 6, 6, 6)?

6.17   Draw a non-Hamiltonian simple graph with degree sequence (3, 3, 3, 6, 6, 6, 9, 9, 9).

Notes

Königsberg was part of East Prussia before Germany's defeat in World War II. It has been renamed as Kaliningrad, and perhaps before long it will get back its original name. It is also the birthplace of German mathematician David Hilbert as well as German philosopher Immanuel Kant. It is interesting to note that even though the Königsberg bridge problem did give birth to Eulerian graphs, Euler himself did not use the concept of Eulerian graphs to solve this problem; instead, he relied on an exhaustive case-by-case verification (see reference [17]).

Ore's theorem (Theorem 6.2.5) can be restated as follows: *If G is a simple graph with $n \geq 3$ vertices and $|N(u)| + |N(v)| \geq n$, for every pair of nonadjacent vertices of G, then G is Hamiltonian.* This statement replaces $d(u)$ in Theorem 6.2.5 by $|N(u)|$. There are several sufficient conditions for a graph to be Hamiltonian using the neighborhood conditions. A nice survey of these results is given in Lesniak [83]. To give a flavor of these results, we give here three results of Faudree, Gould, Jacobson, and Lesniak:

Theorem 1   *If G is a 2-connected graph of order n such that $|N(u) \cap N(v)| \geq \frac{2n-1}{3}$ for each pair u, v of nonadjacent vertices of G, then G is Hamiltonian.*

Theorem 2   *If G is a connected graph of order n such that $|N(u) \cap N(v)| \geq \frac{2n-2}{3}$ for each pair u, v of nonadjacent vertices of G, then G is traceable.*

Theorem 3   *If G is a 3-connected graph of order n such that $|N(u) \cap N(v)| > \frac{2n}{3}$ for each pair u, v of nonadjacent vertices of G, then G is Hamiltonian-connected.*

VII
Graph Colorings

7.0. Introduction

Graph theory would not be what it is today if there had been no coloring problems. In fact, a major portion of the 20th-century research in graph theory has its origin in the four-color problem. (See Chapter VIII for details.)

In this chapter, we present the basic results concerning vertex colorings and edge colorings of graphs. We present two important theorems on graph colorings, namely, Brooks's theorem and Vizing's theorem. We also present a brief discussion on "snarks" and Kirkman's schoolgirls problem.

7.1. Vertex Colorings

Applications of Graph Coloring

We begin with a practical application of graph coloring known as the *storage problem*. Suppose the Department of Chemistry of a college wants to store its chemicals. It is quite probable that some chemicals cause violent reactions when brought together. Such chemicals are *incompatible chemicals*. For safe storage, incompatible chemicals should be kept in distinct rooms. The easiest way to accomplish this is, of course, to store one chemical in each room. But this is certainly not the best way of doing it since we will be using more rooms than are really needed (unless, of course, all the chemicals are mutually incompatible!). So we ask: What is the minimum number of rooms required to store all the chemicals so that in each room only compatible chemicals are stored?

We convert the above storage problem into a problem in graphs. Form a graph $G = G(V, E)$ by making V correspond bijectively to the set of available chemicals and making u adjacent to v if, and only if, the chemicals corresponding to u and v are incompatible. Then, any set of compatible chemical corresponds to a set of independent vertices of G. Thus a safe storing of chemicals corresponds to a partition of V into independent subsets of G. The cardinality of such a minimum partition of V is then the required number of rooms. This minimum cardinality is called the *chromatic number* of the graph G.

Definition 7.1.1   The *chromatic number*, $\chi(G)$, of a graph G is the minimum number of independent subsets that partition the vertex set of G. Any such minimum partition is called a *chromatic partition* of $V(G)$.

The storage problem just described is actually a vertex coloring problem of G. A *vertex coloring* of G is a map $f : V \rightarrow S$, where S is a set of distinct colors; it is *proper* if adjacent vertices of G receive distinct colors of S; that is, if $uv \in E(G)$, then $f(u) \neq f(v)$. Thus $\chi(G)$ is the minimum cardinality of S for which there exists a proper vertex coloring of G by colors of S. Clearly, in any proper vertex coloring of G, the vertices that receive the same color are independent. The vertices that receive a particular color make up a *color class*. Thus, in any chromatic partition of $V(G)$, the parts of the partition constitute the color classes. This allows an equivalent way of defining the chromatic number.

Definition 7.1.2   The *chromatic number* of a graph G is the minimum number of colors needed for a proper vertex coloring of G. G is *k-chromatic*, if $\chi(G) = k$.

Definition 7.1.3   A *k-coloring* of a graph G is a vertex coloring of G that uses k colors.

Definition 7.1.4   A graph G is said to be *k-colorable*, if G admits a *proper* vertex-coloring using k colors.

In considering the chromatic number of a graph, only the adjacency of vertices is taken into account. Hence, multiple edges and loops may be discarded while considering chromatic numbers, unless needed otherwise. As a consequence, we may restrict ourselves to simple graphs when dealing with (vertex) chromatic numbers.

It is clear that $\chi(K_n) = n$. Further, $\chi(G) = 2$ if, and only if, G is bipartite having at least one edge. In particular, $\chi(T) = 2$ for any tree T with at least one edge (since any tree is bipartite). Further (see Figure 7.1),

$$\chi(C_n) = \begin{cases} 2 \text{ if } n \text{ is even} \\ 3 \text{ if } n \text{ is odd.} \end{cases}$$

Exercise 1.1   Prove $\chi(G) = 2$ if, and only if, G is a bipartite graph with at least one edge.

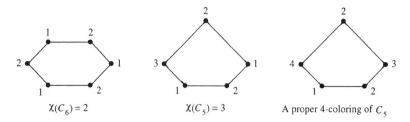

FIGURE 7.1. Illustration of proper vertex coloring

Exercise 1.2 Determine the chromatic number of

(i)   the Petersen graph.
(ii) wheel W_n (see Section 1.7, Chapter I).
(iii) the Herschel graph (see Figure 5.4).
(iv) the Grötzsch graph (see Figure 7.6).

We next consider another application of graph coloring. Let S be a set of students. Each student of S is to take a certain number of examinations for which he or she has registered. Undoubtedly, the examination schedule must be such that all students who have registered for a particular examination will take it at the same time.

Let $\mathbb{P}$ be the set of examinations and for $p \in \mathbb{P}$, let $S(p)$ be the set of students who have to take the examination p. Our aim is to draw up an examination schedule involving only the minimum number of days on the supposition that papers a and b can be given on the same day provided they have no common candidate and that on any day, no candidate shall have more than one examination.

Form a graph $G = G(\mathbb{P}, E)$, where $a, b \in \mathbb{P}$ are adjacent if, and only if, $S(a) \cap S(b) \neq \emptyset$. Then each proper vertex coloring of G yields an examination schedule with the vertices in any color class representing the schedule on a particular day. Thus $\chi(G)$ gives the minimum number of days required for the examination schedule.

Exercise 1.3 Draw up an examination schedule involving the minimum number of days for the following problem:

Sets of students	*Examination subjects*
S_1	algebra, real analysis, and topology
S_2	algebra, operations research, and complex analysis
S_3	real analysis, functional analysis, and complex analysis
S_4	algebra, graph theory, and combinatorics
S_5	combinatorics, topology, and functional analysis
S_6	operations research, graph theory, and coding theory
S_7	operations research, graph theory, and number theory
S_8	algebra, number theory, and coding theory
S_9	algebra, operations research, and real analysis

Exercise 1.4 If G is k-regular, prove that $\chi(G) \geq \frac{n}{n-k}$.

Theorem 7.1.5 gives upper and lower bounds for the chromatic number of a graph G in terms of its independence number and $|V(G)|$.

Theorem 7.1.5   *For any graph G with n vertices and independence number α,*

$$\frac{n}{\alpha} \le \chi \le n - \alpha + 1.$$

Proof   There exists a chromatic partition $\{V_1, V_2, \ldots, V_\chi\}$ of V. Since each V_i is independent $|V_i| \le \alpha$, $1 \le i \le \chi$. Hence $n = \sum_{i=1}^{\chi} |V_i| \le \alpha\chi$, and this gives the inequality on the left.

To prove the inequality on the right, consider a maximum independent set S of α vertices. Then a chromatic partition of $V(G)\backslash S$ together with S would form a partition of $V(G)$ into independent subsets of V. Hence, by definition of χ, $\chi(G) \le \chi(G - S) + 1 \le (n - \alpha) + 1$, since $\chi(G - S) \le |V(G)\backslash S| = (n - \alpha)$.   □

Remark 7.1.6   Unfortunately, none of the above bounds is a good one. For example, if G is the graph obtained by joining C_{2m} with a disjoint $K_{2m}(m \ge 2)$, by an edge, we have $n = 4m$, $\alpha = m + 1$, and $\chi = 2m$, and the above inequalities become $\frac{4m}{m+1} \le 2m \le 3m$.

For a simple graph G, the number $\chi^c = \chi^c(G) = \chi(G^c)$, the chromatic number of G^c is the minimum number of subsets in a partition of $V(G)$ into subsets each inducing a complete subgraph of G.

Bounds on the sum and product of $\chi(G)$ and $\chi^c(G)$ were obtained by Nordhaus and Gaddum [96] (see also reference [61]), as given in Theorem 7.1.7.

Theorem 7.1.7   *For any simple graph G,*

$$2\sqrt{n} \le \chi + \chi^c \le n + 1 \quad and \quad n \le \chi\chi^c \le \left(\frac{n+1}{2}\right)^2.$$

Exercise 1.5   For a simple graph G, prove that $\chi(G^c) \ge \alpha(G)$.

Exercise 1.6   Prove that $\chi(G) \le \ell + 1$, where ℓ is the length of a longest path in G. For each positive integer ℓ, give a graph G with chromatic number $\ell + 1$ and in which any longest path has length ℓ.

Exercise 1.7   Which numbers can be the chromatic numbers of unicyclic graphs? Draw a unicyclic graph on 15 vertices with $\delta = 3$ and having each of these numbers as its chromatic number.

Exercise 1.8   If G is connected and $m \le n$, show that $\chi(G) \le 3$.

7.2. Critical Graphs

Definition 7.2.1   A graph G is called *critical* if for every proper subgraph H of G, $\chi(H) < \chi(G)$. Also, G is *k-critical* if it is k-chromatic and critical.

Remarks 7.2.2
When G is connected, this is equivalent to the condition that $\chi(G-e) < \chi(G)$ for each edge e of G. If $\chi(G) = 1$, then G is either trivial or totally disconnected. Hence, G is 1-critical if, and only if, G is K_1. Again, $\chi(G) = 2$ implies that G is bipartite and has at least one edge. Hence, G is 2-critical if, and only if, G is K_2. For an odd cycle C, $\chi(C) = 3$, and if G contains an odd cycle C properly, G cannot be 3-critical.

Exercise 2.1   Prove that any critical graph is connected.

Exercise 2.2   Prove that for any graph G, $\chi(G-v) = \chi(G)$ or $\chi(G) - 1$ for any $v \in V$, and $\chi(G-e) = \chi(G)$ or $\chi(G) - 1$ for any $e \in E$.

Exercise 2.3   Show that if G is k-critical, $\chi(G-v) = \chi(G-e) = k - 1$ for any $v \in V$ and $e \in E$.

Exercise 2.4   (If $\chi(G-e) < \chi(G)$ for any edge e of G, G is sometimes called *edge-critical*, and if $\chi(G-v) < \chi(G)$ for any vertex v of G, G is called *vertex-critical*). Show that a nontrivial connected graph is vertex-critical if it is edge-critical. Disprove the converse by a counterexample.

Exercise 2.5   Show that a graph is 3-critical if, and only if, it is an odd cycle.

It is clear that any k-chromatic graph contains a k-critical subgraph. (This is seen by removing vertices and edges in succession, whenever possible, without diminishing the chromatic number.)

Theorem 7.2.3   *If G is k-critical, then $\delta \geq k - 1$.*

Proof   Suppose $\delta \leq k - 2$. Let v be a vertex of minimum degree in G. Since G is k-critical, $\chi(G-v) = \chi(G) - 1 = k - 1$ (see Exercise 2.3). Hence, in any proper $(k - 1)$-coloring of $G-v$, at most $(k - 2)$ colors alone would have been used to color the neighbors of v in G. Thus, there is at least one color, say c, that is left out of these $k - 1$ colors. If v is given the color c, a proper $(k - 1)$-coloring of G is obtained. This is impossible since G is k-chromatic. Hence, $\delta \geq (k - 1)$.   □

Corollary 7.2.4   *For any graph, $\chi \leq 1 + \Delta$.*

Proof   Let G be a k-chromatic graph, and let H be a k-critical subgraph of G. Then $\chi(H) = \chi(G) = k$. By Theorem 7.2.3, $\delta(H) \geq k - 1$, and hence $k \leq 1 + \delta(H) \leq 1 + \Delta(H) \leq 1 + \Delta(G)$.   □

Exercise 2.6   Give another proof of Corollary 7.2.4 by using induction on $n = |V(G)|$.

Exercise 2.7   If $\chi(G) = k$, show that G contains at least k vertices each of degree at least $k - 1$.

Exercise 2.8   Prove or disprove: If G is k-chromatic, then G contains a K_k.

Exercise 2.9   Prove: Any $k(\geq 2)$-critical graph contains a $(k - 1)$-critical subgraph.

Exercise 2.10   For each of the graphs G of Exercise 1.2, find a critical subgraph H of G with $\chi(H) = \chi(G)$.

Exercise 2.11   Prove that the wheel $W_{2n-1} = C_{2n-1} \vee K_1$ is a 4-critical graph for each $n \geq 2$. Does a similar statement apply to W_{2n}?

Theorem 7.2.5   *In a critical graph G, no vertex cut is a clique.*

Proof   Suppose G is a k-critical graph and S is a vertex cut of G that is a clique of G (i.e., a complete subgraph of G). Let $H_i, 1 \leq i \leq r$, be the components of $G \backslash S$, and let $G_i = G[V(H_i) \cup S]$. Then each G_i is a proper subgraph of G and hence admits a proper $(k - 1)$-coloring. Since S is a clique, its vertices must receive distinct colors in any proper $(k - 1)$-coloring of G_i. Hence, by fixing the colors for the vertices of S, and coloring, for each i, the remaining vertices of G_i so as to give a proper $(k - 1)$-coloring of G_i, we obtain a proper $(k - 1)$-coloring of G. This contradicts the fact that G is k-chromatic. (See Figure 7.2.)   □

Corollary 7.2.6   *Every critical graph is a block.*

Exercise 2.12*   Prove that every k-critical graph is $(k - 1)$-edge connected (G. A. Dirac [34]).

Exercise 2.13   Show by means of an example that criticality is essential in Exercise 2.12, that is, a k-chromatic graph need not be $(k - 1)$-edge connected.

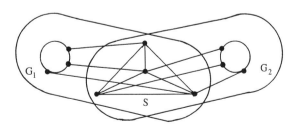

FIGURE 7.2. $G[S] \cong K_4$ $(r = 2)$

Brooks's Theorem

We next consider *Brooks's [22] theorem*. Recall Corollary 7.2.4, which states that $\chi \leq 1 + \Delta$. If G is an odd cycle, $\chi(G) = 3 = 1 + 2 = 1 + \Delta(G)$, whereas if G is a complete graph, say K_k, $\chi(G) = k = 1 + (k - 1) = 1 + \Delta(G)$. That these are the only extremal families of graphs for which $\chi(G) = 1 + \Delta(G)$ is the assertion of Brooks's theorem.

Theorem 7.2.7 (Brooks's theorem)   *If a connected graph G is neither an odd cycle nor a complete graph, then $\chi(G) \leq \Delta(G)$.*

Proof   If $\Delta(G) \leq 2$, then G is either a path or a cycle. For a path G (other than K_1 and K_2), and for an even cycle G, $\chi(G) = 2 = \Delta(G)$. According to our assumption G is not an odd cycle. So let $\Delta(G) \geq 3$.

The proof is by contradiction. Suppose the result is not true. Then there exists a minimal graph G of maximum degree $\Delta(G) = \Delta \geq 3$ such that G is not Δ-colorable, but for any vertex v of G, $G{-}v$ is $\Delta(G{-}v)$-colorable and therefore Δ-colorable.

Claim 1. Let v be any vertex of G. Then in any proper Δ-coloring of $G{-}v$, all the Δ-colors must be used for coloring the neighbors of v in G. Otherwise, if some color i is not represented in $N_G(v)$, then v could be colored using i, and this would give a Δ-coloring of G, a contradiction to the choice of G. Thus, G is a Δ-regular graph satisfying claim 1.

For $v \in V(G)$, let $N(v) = \{v_1, v_2, \ldots, v_\Delta\}$. In a proper Δ-coloring of $G{-}v = H$, let v_i receive color i, $1 \leq i \leq \Delta$. For $i \neq j$, let H_{ij} be the subgraph of H induced by the vertices receiving the ith and jth colors.

Claim 2. v_i and v_j belong to the same component of H_{ij}. Otherwise, the colors i and j can be interchanged in the component of H_{ij} that contains the vertex v_j. Such an interchange of colors once again yields a proper Δ-coloring of H. In this new coloring, both v_i and v_j receive the same color, namely, i, a contradiction to claim 1. This proves claim 2.

Claim 3. If C_{ij} is the component of H_{ij} containing v_i and v_j, then C_{ij} is a path in H_{ij}. As before, $N_H(v_i)$ contains exactly one vertex of color j. Further, C_{ij} cannot contain a vertex, say y, of degree at least 3; for, if y is the first such vertex on a $v_i{-}v_j$ path in C_{ij} that has been colored, say, with i, then at least three neighbors of y in C_{ij} have the color j. Hence, we can recolor y in H with a color different from both i and j, and in this new coloring of H, v_i and v_j would belong to distinct components of H_{ij} (see Figure 7.3 (a)). (Note that by our choice of y, any $v_i{-}v_j$ path in H_{ij} must contain y.) But this contradicts claim 2.

Claim 4. $C_{ij} \cap C_{ik} = \{v_i\}$ for $j \neq k$. Indeed, if $w \in C_{ij} \cap C_{ik}$, $w \neq v_i$, then w is adjacent to two vertices of color j on C_{ij} and two vertices of color k on C_{ik} (Figure 7.3 (b)). Again, we can recolor w in H by giving a color different from the colors of the neighbors of w in H. In this new coloring of H, v_i and v_j belong to distinct components of H_{ij}, a contradiction to claim 1. This completes the proof of claim 4.

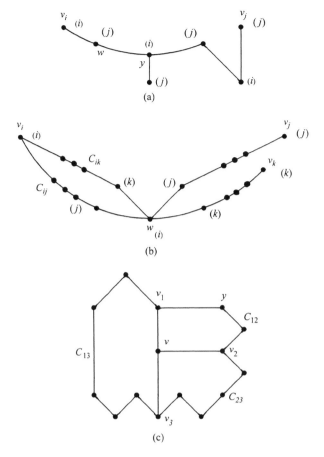

FIGURE 7.3. Graphs for proof of Theorem 7.2.7 (The numbers inside the parentheses denote the vertex colors.)

We are now in a position to complete the proof of the theorem. By hypothesis, G is not complete. Hence, G has a vertex v, and a pair of nonadjacent vertices v_1 and v_2 in $N_G(v)$ (see Exercise 4.11, Chapter I). Then the v_1–v_2 path C_{12} in H_{12} of $H = G - v$ contains a vertex $y(\neq v_2)$ adjacent to v_1. Naturally, y would receive color 2. Since $\Delta \geq 3$, there exists a vertex $v_3 \in N_G(v)$. Now interchange colors 1 and 3 in the path C_{13} of H_{13}. This would result in a new coloring of $H = G - v$. Denote the v_i–v_j path in H under this new coloring by C'_{ij} (See Figure 7.3(c)). Then $y \in C'_{23}$, since v_1 receives color 3 in the new coloring (whereas y retains color 2). Also, $y \in C_{12} - v_1 \subset C'_{12}$. Thus, $y \in C'_{23} \cap C'_{12}$. This contradicts claim 4 (since $y \neq v_2$) and the proof is complete.   □

Other Coloring Parameters

There are several other vertex coloring parameters of a graph G. We mention two of these next. Let f be a k-coloring (not necessarily proper) of G, and let $(V_1, \ldots, V_k)$

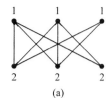
(a)

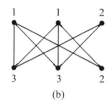
(b)

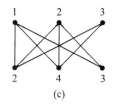
(c)

FIGURE 7.4. Different colorings of $K_{3,3}-e$

be the color classes of $V(G)$ induced by f. Coloring f is *pseudocomplete* if between any two distinct color classes, there is at least one edge of G. f is *complete* if it is pseudocomplete and each V_i, $1 \le i \le k$, is an independent set of G. Thus, $\chi(G)$ is the minimum k for which G has a complete k-coloring f.

Definition 7.2.8   The *achromatic number* $\phi(G)$ of a graph G is the maximum k for which G has a complete k-coloring.

Definition 7.2.9   The *pseudoachromatic number* $\psi(G)$ of G is the maximum k for which G has a pseudocomplete k-coloring.

Example 7.2.10   Figure 7.4 gives (a) chromatic, (b) achromatic, and (c) pseudoachromatic colorings of $K_{3,3}-e$.
It is clear that for any graph G, $\chi(G) \le \phi(G) \le \psi(G)$.

Exercise 2.14   Let G be a graph and H a subgraph of G. Prove that $\chi(H) \le \chi(G)$ and $\psi(H) \le \psi(G)$. Show by means of an example that $\phi(H) \le \phi(G)$ need not always be true.

Exercise 2.15   Prove:
   (i) $\psi(\psi - 1) \le 2m$.
   (ii) $\psi(K_a \vee K_b^c) = a + 1$.

From (ii), deduce that for any graph, $\psi \le n - \alpha + 1$.
A description of several other coloring parameters can be found in Jensen and Toft [72].

7.3.   Triangle-Free Graphs

Definition 7.3.1   A graph G is *triangle-free* if G contains no K_3.

Remark 7.3.2   Triangle-free graphs cannot contain a K_k, $k \ge 3$, either. It is obvious that if G contains a clique of size k, then $\chi(G) \ge k$. However, the converse is not true. That is, if G has a large chromatic number, then G need not contain a

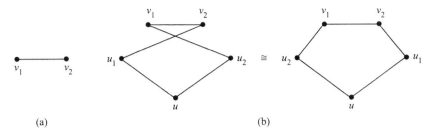

FIGURE 7.5. Graph for proof of Theorem 7.3.3

clique of large size. In fact, one can construct a graph even without triangles but with a prescribed chromatic number.

Theorem 7.3.3 (Myceilski [92]) *For every positive integer k, there exists a triangle-free graph with chromatic number k.*

Proof For $k = 1$ and $k = 2$, the graphs K_1 and K_2 satisfy the conditions of the theorem. For $k = 3$, the cycle C_5 has the desired property; it is triangle-free and 3-chromatic. The transition from K_2 in the case $k = 2$ to the graph C_5 in the case $k = 3$ is depicted in Figure 7.5.

The graph in Figure 7.5(b) is obtained from $K_2 = (v_1 v_2)$ by taking two new independent vertices $u_i, i = 1, 2$, and making u_i adjacent to all the neighbors of v_i in K_2 (here, making u_1 adjacent to v_2 and u_2 adjacent to v_1), and taking a new vertex u and making it adjacent to all the u_i's.

The result is now established by generalizing the above construction. Suppose a triangle-free k-chromatic graph G_k has already been constructed with $V(G_k) = \{v_1, v_2, \ldots, v_n\}$. Form a supergraph G_{k+1} of G_k by taking a new set $\{u_1, u_2, \ldots, u_n\}$ of independent vertices and making u_i adjacent in G_{k+1} to all the neighbors of v_i in G_k, $1 \leq i \leq n$. In addition, another new vertex u is taken and made adjacent to all the u_i's. Since G_3 is C_5, this construction yields graph G_4, which is the Grötszch graph (see Figure 7.6).

We now show that the graph G_{k+1} is triangle-free. Since G_k is triangle-free, should G_{k+1} contain a triangle, it can be only of the form $u_i v_j v_k u_i$ (since $\{u_1, u_2, \ldots, u_n\}$ is an independent subset of G_{k+1} and u is nonadjacent to $v_1, v_2, \ldots, v_n$). Since u_i is adjacent only to the neighbors of v_i, then v_j and v_k must be neighbors of v_i in G_k. Thus, $v_i v_j v_k v_i$ is a triangle in G_k, a contradiction.

Next, we show that G_{k+1} is $(k + 1)$-chromatic. We note that G_{k+1} is $(k + 1)$-colorable. This follows from the fact that G_k is k-colorable. Starting with a proper k-coloring of G_k, we can assign to u_i the color already assigned to v_i, $1 \leq i \leq n$, and to u the $(k + 1)$th color. This yields a proper $(k + 1)$-coloring for G_{k+1}.

Suppose $\chi(G_{k+1}) < k + 1$. Then G_{k+1} is k-colorable. In a proper k-coloring of G_{k+1}, u receives a certain color, say the first color. Then $u_1, u_2, \ldots, u_n$ are colored

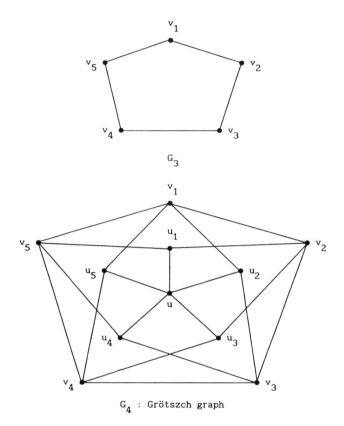

$$G_3$$

$$G_4 \text{ : Grötszch graph}$$

FIGURE 7.6. Graph G_3 and the Grötszch graph, G_4

by the remaining $(k-1)$ colors. Now recolor v_i by the color of u_i, $1 \le i \le n$. This gives a proper $(k-1)$-coloring for G_k, a contradiction to our choice that $\chi(G_k) = k$. Thus, G_{k+1} has all the desired properties.   □

Remark 7.3.4   Observe that the graph G_k described in the above proof is k-critical. Thus, for each $k \ge 1$, there exists a k-critical triangle-free graph. Not every k-critical graph is triangle-free; for example, the complete graph $K_k(k \ge 3)$ is k-critical, but is not triangle-free.

7.4.   Edge Colorings of Graphs

The Timetable Problem

Suppose in a school there are r teachers, $T_1, T_2, \ldots, T_r$, and s classes, $C_1, C_2, \ldots, C_s$. Each teacher T_i is expected to teach the class C_j for p_{ij} periods. It is clear that during any particular period, no more than one teacher can handle a particular class and no more than one class can be engaged by any teacher. Our aim is to draw up a timetable for the day that requires only the minimum number of periods. This problem is known as the "timetable problem."

To convert this problem into a graph-theoretic one, we form the bipartite graph $G = G(T, C)$ with bipartition (T, C), where T represents the set of teachers T_i and C represents the set of classes C_j. Further, T_i is made adjacent to C_j in G with p_{ij} edges if, and only if, teacher T_i is to handle class C_j for p_{ij} periods. Now, color the edges of G so that no two adjacent edges receive the same color. Then the edges in a particular color class, that is, the edges in that color form a matching in G and correspond to a schedule of work for a particular period. Hence, the minimum number of periods required is the minimum number of colors in an edge-coloring of G in which adjacent edges receive distinct colors; in other words, it is the edge-chromatic number of G. We now present these notions as formal definitions. (For a more detailed description of the timetable problem, see Chapter X.)

Definition 7.4.1 An *edge-coloring* of a loopless graph G is a function $\pi : E(G) \rightarrow S$, where S is a set of distinct colors; it is *proper* if no two adjacent edges receive the same color. Thus a proper edge-coloring π of G is a function $\pi : E(G) \rightarrow S$ such that $\pi(e) \neq \pi(e')$ whenever edges e and e' are adjacent in G.

Definition 7.4.2 The minimum k for which a loopless graph G has a proper k-edge-coloring is called the *edge chromatic number* or *chromatic index* of G. It is denoted by $\chi'(G)$. G is k-*edge-chromatic* if $\chi'(G) = k$.

Further, if an edge uv is colored by color c, we say that c is represented at both u and v. If G has a proper k-edge-coloring, $E(G)$ is partitioned into k edge-disjoint matchings.

It is clear that for any (loopless) graph G, $\chi'(G) \geq \Delta(G)$ since the $\Delta(G)$ edges incident at a vertex v of maximum degree $\Delta(G)$ must all receive distinct colors. For bipartite graphs, however, equality holds.

Theorem 7.4.3 (König)   *If G is a loopless bipartite graph, $\chi'(G) = \Delta(G)$.*

Proof   The proof is by induction on the size (i.e., number of edges) m of G. The result is true for $m = 1$. Assume the result for bipartite graphs of size at most $m - 1$. Let G have m edges. Let $e = uv \in E(G)$. Then $G-e$ has (since $\Delta(G-e) \leq \Delta(G)$) a proper Δ-edge-coloring, say c. Out of these Δ colors, suppose that, one particular color is not represented at both u and v. Then edge uv can be colored with this color and a proper Δ-edge-coloring of G is obtained.

In the other case (i.e., in the case for which each of the Δ colors is represented either at u or at v), since the degrees of u and v in $G-e$ are at most $\Delta - 1$, there exists a color out of the Δ colors that is not represented at u, and similarly there exists a color not represented at v. Thus, if color j is not represented at u in c, then j is represented at v in c, and if color i is not represented at v in c, then i is represented at u in c. Since G is bipartite and u and v are not in the same parts of the bipartition, there can exist no u–v path in G in which the colors alternate between i and j.

Let P be a maximal path in $G-e$ starting from u in which the colors of the edges alternate between i and j. Interchange the colors i and j in P. This would still yield an edge-coloring of $G-e$ using the Δ colors in which color i is not represented at both u and v. Now, color the edge uv by the color i. This results in a proper Δ-edge-coloring of G. □

Exercise 4.1 Disprove the converse of Theorem 7.4.3 by a counterexample.

Next, we determine the chromatic index of the complete graphs.

Theorem 7.4.4 $\chi'(K_n) = \begin{cases} n-1 \ \text{if } n \text{ is even,} \\ n \ \text{if } n \text{ is odd.} \end{cases}$

Proof (Berge) Since K_n is regular of degree $n-1$, $\chi'(K_n) \geq n-1$.

Case 1. n is even. We now show that $\chi'(K_n) \leq n-1$ by exhibiting a proper $(n-1)$-edge-coloring of K_n. Label the n vertices of K_n as $0, 1, \ldots, n-1$. Draw a circle with center at 0 and place the remaining $n-1$ numbers on the circumference of the circle so that they form a regular $(n-1)$-gon (Figure 7.7). Then the $\frac{n}{2}$ edges $(0,1), (2, n-1), (3, n-2), \ldots, (\frac{n}{2}, \frac{n}{2}+1)$ form a 1-factor of K_n. These $\frac{n}{2}$ edges

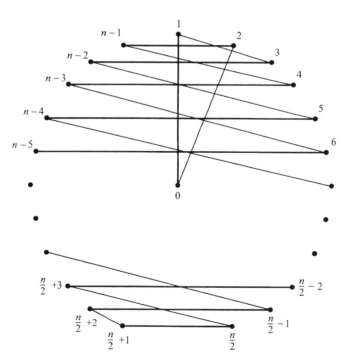

FIGURE 7.7. Graph for proof of Theorem 7.4.4

are the thick edges of Figure 7.7. Rotation of these edges through the angle $\frac{2\pi}{n-1}$ in succession gives $(n-1)$-edge-disjoint 1-factors of K_n. This would account for $\frac{n}{2}(n-1)$ edges, or all the edges of K_n. (Actually, the above construction displays a 1-factorization of K_n when n is even.) Each 1-factor can be assigned a color. Thus $\chi'(K_n) \leq n-1$. This proves the result in case 1.

Case 2. n is odd. Take a new vertex and make it adjacent to all the n vertices of K_n. This gives K_{n+1}. By case 1, $\chi'(K_{n+1}) = n$. The restriction of this edge coloring to K_n yields a proper n-edge-coloring of K_n. Hence, $\chi'(K_n) \leq n$. However, K_n cannot be edge-colored properly with $n-1$ colors. This is because the size of any matching of K_n can contain no more than $\frac{n-1}{2}$ edges and hence $n-1$ matchings of K_n can contain no more than $\frac{(n-1)^2}{2}$ edges. But K_n has $\frac{n(n-1)}{2}$ edges. Thus, $\chi'(K_n) \geq n$, and hence $\chi'(K_n) = n$. □

Exercise 4.2 Show that a Hamiltonian cubic graph is 3-edge-chromatic.

Exercise 4.3 Show that the Petersen graph is 4-edge-chromatic.

Exercise 4.4 Show that the Herschel graph (see Figure 5.4) is 4-edge-chromatic.

Exercise 4.5 Determine the edge-chromatic number of the Grötszch graph (Figure 7.6).

Exercise 4.6 Show that a simple cubic graph with a cut edge is 4-edge-chromatic.

Exercise 4.7 Describe a proper k-edge-coloring of a k-regular bipartite graph.

Exercise 4.8 Show that any bipartite graph G of maximum degree Δ is a subgraph of a Δ-regular bipartite graph. Hence, furnish an alternative proof of Theorem 7.4.3, using Exercise 4.7.

Vizing's Theorem

Although it is true that for any loopless graph G, $\chi'(G) \geq \Delta(G)$, it turns out that for any simple graph G, $\chi'(G) \leq 1 + \Delta(G)$. This major result in edge-coloring of graphs was established by Vizing [118] and independently by Gupta [56].

Theorem 7.4.5 (Vizing–Gupta) *For any simple graph G, $\Delta(G) \leq \chi'(G) \leq 1 + \Delta(G)$.*

Proof In a proper edge-coloring of G, $\Delta(G)$ colors are to be used for the edges incident at a vertex of maximum degree in G. Hence, $\chi'(G) \geq \Delta(G)$.

We now prove that $\chi'(G) \leq 1 + \Delta$, where $\Delta = \Delta(G)$.

If G is not $(1+\Delta)$-edge-colorable, choose a subgraph H of G with maximum possible number of edges such that H is $(1+\Delta)$-edge-colorable. We derive a

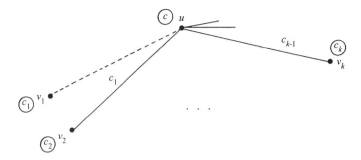

FIGURE 7.8. Graph for proof of Theorem 7.4.5

contradiction by showing that there exists a subgraph H_0 of G that is $(1+\Delta)$-edge-colorable and has one edge more than H.

By our assumption, G has an edge $uv_1 \notin E(H)$. Since deg $u \le \Delta$, and $1 + \Delta$ colors are being used in H, there is a color c that is not represented at u (i.e., not used for any edge of H incident at u). For the same reason, there is a color c_1 not represented at v_1. (See Figure 7.8, where the color not represented at a particular vertex is enclosed in a circle and marked near the vertex.)

There must be an edge, say uv_2, colored c_1; otherwise, uv_1 can be assigned the color c_1, and $H \cup (uv_1)$, which has one edge more than H, would have a proper $(1 + \Delta)$-edge-coloring. Again there is a color, say c_2, not represented at v_2. Then, as above, there is an edge uv_3 colored c_2 and there is a color, say c_3, not represented at v_3.

In this way, we construct a sequence of edges $\{uv_1, uv_2, \ldots, uv_k\}$ such that color c_i is not represented at vertex v_i, $1 \le i \le k$, and the edge uv_{j+1} receives the color c_j, $1 \le j \le k - 1$ (see Figure 7.8).

Suppose at some stage, say the mth stage, where $1 \le m \le k$, c (the missing color at u) is not represented at v_m. We then "cascade" (i.e., shift in order) the colors $c_1, \ldots, c_{m-1}$ from $uv_2, uv_3, \ldots, uv_m$ to $uv_1, uv_2, \ldots, uv_{m-1}$. Under this new coloring, c is not represented both at u and v_m, and therefore we can color uv_m with c. This yields a proper $(1 + \Delta)$-edge-coloring to $H \cup (uv_1)$, contradicting the choice of H.

Hence, we may assume that c is represented at each of the vertices $v_1, v_2, \ldots, v_k$.

Now we need to know why the sequence of edges, uv_i, $1 \le i \le k$, had stopped. There are two possible reasons. Either there is no edge incident to u that is colored c_k, or the color $c_k = c_j$ for some $j < k - 1$ and so has already been represented at u. Note that the sequence must stop at some finite stage since $d(u)$ is finite; however, it may as well stop before all the edges incident to u are exhausted.

If c_k is not represented at u in H, then we can cascade as before so that uv_i gets color c_i, $1 \le i \le k - 1$, and then color uv_k with color c_k. Once again, we have a contradiction to our assumption on H.

Thus, we must have $c_k = c_j$ for some $j < k - 1$. In this case, cascade the colors $c_1, c_2, \ldots, c_j$ so that uv_i has color c_i, $1 \le i \le j$, and leave uv_{j+1} uncolored

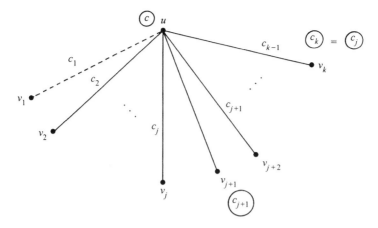

FIGURE 7.9.   Another graph for proof of Theorem 7.4.5

(Figure 7.9). Let $S = (H \cup (uv_1)) - uv_{j+1}$. Then S and H have the same number of edges.

Now consider S_{cc_j}, the subgraph of S defined by the edges of S with colors c and c_j. Clearly, each component of S_{cc_j} is either an even cycle or a path in which the adjacent edges alternate with colors c and c_j.

Now, c is represented at each of the vertices $v_1, v_2, \ldots, v_k$, and in particular at v_{j+1} and v_k. But c_j is not represented at v_{j+1} and v_k, since we have just moved c_j to uv_j, and $c_j = c_k$ is not represented at v_k. Hence in S_{cc_j}, the degrees of v_{j+1} and v_k are both equal to one. Moreover, c_j is represented at u but c is not. Therefore, u also has degree one in S_{cc_j}. As each component of S_{cc_j} is either a path or an even cycle, not all of u, v_{j+1}, and v_k can be in the same component of S_{cc_j} (since a path can contain only two vertices of degree 1).

If u and v_{j+1} are in different components of S_{cc_j}, then interchange the colors c and c_{j+1} in the component containing v_{j+1}. Then c is not represented at both u and v_{j+1}, and so we can color the edge uv_{j+1} with c. This gives a $(1+\Delta)$-edge-coloring to the graph $S \cup (uv_{j+1})$.

Suppose, then, that u and v_{j+1} are in the same component of S_{cc_j}. Then, necessarily, v_k is not in this component. Interchange c and c_j in the component containing v_k. In this case, *further* cascade the colors so that uv_i has color c_i, $1 \le i \le k - 1$. Now color uv_k with color c.

Thus, we have extended our edge coloring of S with $1 + \Delta$ colors to one more edge of G. This contradiction proves that $H = G$, and thus $\chi'(G) \le 1 + \Delta$. $\square$

Actually, Vizing proved a more general result than the one given above. Let G be any loopless graph and let μ denote the maximum number of edges joining two vertices in G. Then the generalized Vizing's theorem states that $\Delta \le \chi' \le \Delta + \mu$. This theorem is the best possible in that there are graphs with $\chi' = \Delta + \mu$. For example, let G be the graph of Figure 7.10. Since any two edges of G are adjacent,

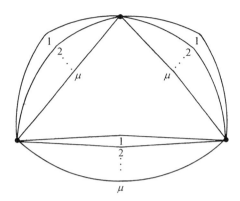

FIGURE 7.10. Graph illustrating the generalized Vizing's theorem

$\chi' = m(G) = 3\mu = \Delta + \mu$. For a proof of the generalized Vizing's theorem, see Yap [122].

Definition 7.4.6 Graphs for which $\chi' = \Delta$ are called *class 1 graphs*, and those for which $\chi' = 1 + \Delta$ are called *class 2 graphs*.

Example 7.4.7 Bipartite graphs are of class 1 (see Theorem 7.4.3), whereas the Petersen graph (see Exercise 4.3) and any simple cubic graph with a cut edge (see Exercise 4.6) are of class 2. For details relating to graphs of class 1 and class 2, see references [39] and [122].

Exercise 4.9 Let G be a simple Δ-edge-chromatic critical graph (i.e., G is of class 1 and for every edge e of G, $\chi'(G-e) < \chi'(G)$). Prove that if $uv \in E(G)$, then $d(u) + d(v) \geq \Delta + 2$.

7.5. Snarks

A consequence of the Vizing-Gupta theorem is that *if, G is a cubic simple graph, then $\chi'(G) = 3$ or 4*. By Exercise 4.6, *if G is a cubic simple graph with a cut edge, then $\chi'(G) = 4$*. So the natural question is: Are there 2-edge connected simple cubic graphs that are 4-edge-chromatic? Such graphs are important in their own right, since their existence is related to the four-color problem (see Chapter VIII). The search for such graphs has led to the study of snarks.

Definition 7.5.1 A *snark* is a cyclically 4-edge connected cubic graph of girth at least 5 that has chromatic index 4.

Exercise 7.1 Prove that no snark can be Hamiltonian.

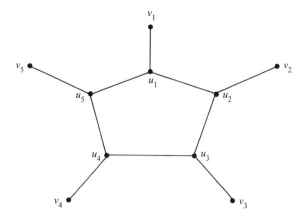

FIGURE 7.11. Graph for proof of Theorem 7.5.2

Clearly, the Petersen graph is a snark. In fact, Theorem 7.5.2 is an interesting result.

Theorem 7.5.2   *The Petersen graph P is the smallest snark and it is the unique snark on 10 vertices.*

Proof   Let G be a snark, and A a cyclical edge cut of G. Then $G-A$ has two components, each having a cycle of length at least 5 (since G is of girth at least 5). Hence, $|V(G)| \geq 10$. Thus, P is a smallest snark, since $|V(G)| = 10$.

We now show that any snark G on 10 vertices must be isomorphic to P. Let A be a cyclical edge-cut of G with $|A| = 4$. Then each component of $G-A$ is a 5-cycle. But this will not account for all 15 edges of G. If $|A| > 5$, then $|E(G)| > 5+5+5 = 15$, a contradiction. Hence, $|A| = 5$, and let $A = \{u_i v_i : 1 \leq i \leq 5\}$. Then $G-A$ consists of two 5-cycles. Let one of these cycles be $\{u_1, u_2, u_3, u_4, u_5\}$. Let v_i be the third neighbor of u_i not belonging to the set $\{u_1, u_2, u_3, u_4, u_5\}$ for each i. If $v_1 v_2$ or $v_1 v_5$ is an edge of G, then G contains a 4-cycle (see Figure 7.11).

Since G is cubic, $v_1 v_3 \in E(G)$ and $v_1 v_4 \in E(G)$. Similarly, $v_2 v_4$, $v_2 v_5$, and $v_3 v_5$ are edges of G, that is, $G \cong P$.   □

The construction of snarks is not easy. In 1975, Isaacs constructed two infinite classes of snarks. Prior to that, only four kinds of snarks were known: (1) the Petersen graph on 10 vertices, (2) Blanuša's graphs on 18 vertices, (3) Szekeres's graph on 50 vertices, and (4) Blanche Descartes's graph on 210 vertices.

7.6.   Kirkman's Schoolgirls Problem

Kirkman's schoolgirls problem was introduced in 1850 by Reverend Thomas J. Kirkman as "query 6" in page 48 of the "Ladies and Gentlemen's Diary." The problem is the following: A teacher would like to take 15 schoolgirls out for a walk, the girls being arranged in five rows of three. The teacher would like to ensure equal chances of friendship between any two girls. Hence, it is desirable to find different row arrangements for the seven days of the week such that any pair of girls walk in the same row on exactly one day of the week.

Kirkman's fifteen schoolgirls problem has a solution. In fact, the following is one of the possible schedules:

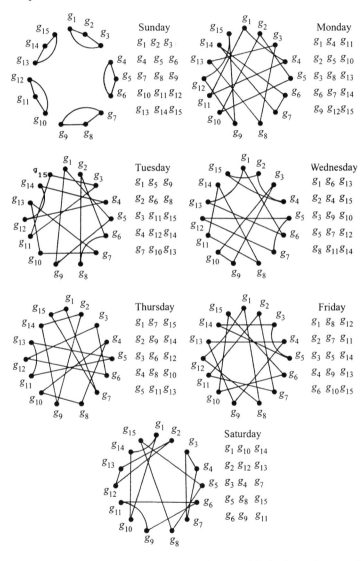

In the general case, one wants to arrange $6n + 3$ girls in $2n + 1$ rows of three. The problem is to find different row arrangements for $3n + 1$ different days in such a way that any pair of girls walks in the same row on exactly one day out of the $3n + 1$ days. The existence of such an arrangement was proved by Ray-Chaudhuri and R. M. Wilson [106]. In graph theoretic terminology, Kirkman's schoolgirls problem corresponds to an edge-coloring $\mathcal{C} : E(K_{6n+3}) \rightarrow \{c_1, \ldots, c_{3n+1}\}$ of the complete graph $G = K_{6n+3}$ with $3n + 1$ colors such that if E_i denotes the set of all edges receiving the color c_i, and $G_i = G[E_i]$, then G_i has $2n + 1$ components, each component being a triangle.

The general problem can be tackled as follows: Consider the triangle graph T of K_{6n+3}. T is defined as follows: The vertex set of T is the set of all triads of $V(K_{6n+3})$, and two distinct vertices of T are joined by an edge in T if, and only if, the corresponding triads have two elements in common. Let S be any independent set of T. Each vertex of S gives rise to three pairs of vertices of K_{6n+3}, and each such pair belongs to at most one vertex of S. Hence, we have $3|S| \leq \binom{6n+3}{2}$, that is, $|S| \leq (2n + 1)(3n + 1)$. We must then find an independent set S' of cardinality $|S'| = (2n + 1)(3n + 1)$. Such a set exists, since every solution of the Kirkman's schoolgirls problem yields an independent set of T with $(2n + 1)(3n + 1)$ vertices. We observe that S' covers each pair of $V(K_{6n+3})$ exactly once. Having found a maximum independent set S' in T, we form a new graph T' as follows: We take S' as its vertex set and join two vertices of T' by an edge if, and only if, the corresponding triads have exactly one vertex in common. We note that each independent set of T' is a partition of a subset of $V(K_{6n+3})$ into subsets of cardinality three, and hence each independent set of T' has at most $(2n + 1)$ vertices. If the chromatic number of T' is $3n + 1$, then there is a partition $(V_1, V_2, \ldots, V_{3n+1})$ of $V(T')$ into parts of size $2n + 1$ each. This partition is a solution to the Kirkman's schoolgirls problem, and conversely, each solution to the Kirkman's schoolgirls problem yields such a partition.

Exercise 6.1 Let n and m be such that there exist edge partitions $\mathcal{F}$ and $\mathcal{G}$ of K_n and K_m, respectively, into triangles with $\mathcal{F} \subset \mathcal{G}$. Prove that $m \geq 2n + 1$.

7.7. Chromatic Polynomials

In 1946, Birkhoff and Lewis [16] introduced the chromatic polynomial of a graph in their attempt to tackle the four-color problem (see Chapter VIII) through algebraic techniques.

For a graph G and a given set of λ colors, the function $f(G; \lambda)$ is defined to be the number of ways of (vertex) coloring G properly using the λ colors. Hence, $f(G; \lambda) = 0$ when G has no proper λ-coloring. Clearly, the minimum λ for which $f(G; \lambda) > 0$ is the chromatic number $\chi(G)$ of G.

It is easy to see that $f(K_n; \lambda) = \lambda(\lambda - 1) \ldots (\lambda - n + 1)$ for $\lambda \geq n$. This is because any vertex of K_n can be colored by any of the given λ colors. After coloring a vertex of K_n, a second vertex of K_n can be colored by any one of the remaining $(\lambda - 1)$ colors, and so on. In particular, $f(K_3; \lambda) = \lambda(\lambda - 1)(\lambda - 2)$. Also, $f(K_n^c; \lambda) = \lambda^n$.

Let $e = uv$ be any edge of G. Recall (see Section 4.3, Chapter IV) that the graph $G.e$ is obtained from G by contracting the edge e. Theorem 7.7.1 presents a simple reduction formula to compute $f(G; \lambda)$.

Theorem 7.7.1 *Let G be any graph. Then $f(G; \lambda) = f(G-e; \lambda) - f(G \cdot e; \lambda)$ for any edge e of G.*

Proof $f(G-e; \lambda)$ denotes the number of proper colorings of $G-e$ using λ colors. Hence, it is the sum of the number of proper colorings of $G-e$ in which u and v receive the same color and the number of proper colorings of $G-e$ in which u and v receive distinct colors. The former number is $f(G.e; \lambda)$, and the latter number is $f(G; \lambda)$. □

Exercise 7.1 If G and H are disjoint, show that

$$f(G \cup H; \lambda) = f(G; \lambda)f(H; \lambda).$$

Theorem 7.7.1 could be used recursively to determine $f(G; \lambda)$ for graphs of small size by taking the given graph on n vertices as G and successively deleting edges until we end up with the totally disconnected graph K_n^c. It can also be determined by taking the given graph as $G-e$ and recursively adding new edges e until we end up with the complete graph K_n. For a fixed n, when $m(G)$, the number of edges of G, is small, the first method is preferable, and when it is large, the second method is preferable. These two methods are illustrated for the graph C_4. (Here the function $f(G; \lambda)$ is represented by the graph itself.)

Method 1

$$= (\lambda(\lambda - 1))^2 - \{\lambda^2(\lambda - 1) - \lambda(\lambda - 1)\} - \lambda(\lambda - 1)(\lambda - 2)$$
$$= \lambda^4 - 4\lambda^3 + 6\lambda^2 - 3\lambda$$

Therefore, $f(C_4; \lambda) = \lambda^4 - 4\lambda^3 + 6\lambda^2 - 3\lambda$.

Method 2

$$f(C_4; \lambda) = \left(\begin{array}{c} \square \\ G-e \end{array} \right)$$

$$= \left(\begin{array}{c} \boxed{e} \\ G \end{array} \right) + \left(\begin{array}{c} \infty \\ G \cdot e \end{array} \right)$$

$$= \left(\begin{array}{c} \boxed{} \end{array} \right) + \left(\begin{array}{c} \vdots \end{array} \right)$$

$$= \left(\begin{array}{c} \boxtimes \end{array} \right) + \left(\begin{array}{c} \triangleright \end{array} \right) + \left(\begin{array}{c} \vdots \end{array} \right) + \left(\begin{array}{c} \circ \end{array} \right)$$

$$= f(K_4; \lambda) + f(K_3; \lambda) + f(K_3; \lambda) + f(K_2; \lambda)$$
$$= f(K_4; \lambda) + 2f(K_3; \lambda) + f(K_2; \lambda)$$
$$= \lambda(\lambda - 1)(\lambda - 2)(\lambda - 3) + 2\lambda(\lambda - 1)(\lambda - 2) + \lambda(\lambda - 1)$$
$$= \lambda^4 - 4\lambda^3 + 6\lambda^2 - 3\lambda.$$

The function $f(C_4; \lambda)$ computed above is a monic polynomial with integer coefficients of degree $n = 4$ in which the coefficient of $\lambda^3 = -4 = -m$, the constant term is zero, and the coefficients alternate in sign. That this is the case with all such functions $f(G; \lambda)$ is the content of Theorem 7.7.2. For this reason, $f(G; \lambda)$ is called the *chromatic polynomial* of the graph G.

Theorem 7.7.2   *For a simple graph G of order n and size m, $f(G; \lambda)$ is a monic polynomial of degree n in λ with integer coefficients and constant term zero. In addition, its coefficients alternate in sign and the coefficient of λ^{n-1} is $-m$.*

Proof   Proof is by induction on m. If $m = 0$, G is K_n^c and $f(K_n^c; \lambda) = \lambda^n$, and the statement of the theorem is trivially true in this case. Suppose, now, that the theorem holds for all graphs with fewer than m edges, where $m \geq 1$. Let G be any simple graph of order n and size m, and let e be any edge of G. Both $G-e$ and $G.e$ (after removal of multiple edges, if necessary) are simple graphs with at most $(m - 1)$ edges, and hence, by induction hypothesis,

$$f(G-e; \lambda) = \lambda^n - a_0 \lambda^{n-1} + a_1 \lambda^{n-2} + \cdots + (-1)^{n-1} a_{n-2}\lambda,$$

and

$$f(G.e; \lambda) = \lambda^{n-1} - b_1\lambda^{n-2} + \cdots + (-1)^{n-2}b_{n-2}\lambda,$$

where $a_0, \ldots, a_{n-1}; b_1, \ldots, b_{n-2}$ are nonnegative integers (so that the coefficients alternate in sign), and a_0 is the number of edges in $G-e$, which is $m-1$. By Theorem

7.7.1, $f(G; \lambda) = f G{-}e; \lambda) - f(G.e; \lambda)$, and hence $f(G; \lambda) = \lambda^n - (a_0 + 1)\lambda^{n-1} + (a_1 + b_1)\lambda^{n-2} - \ldots + (-1)^{n-1}(a_{n-2} + b_{n-2})\lambda$. Since $a_0 + 1 = m$, $f(G; \lambda)$ has all the stated properties. $\square$

Theorem 7.7.3 *A simple graph G on n vertices is a tree if, and only if,* $f(G; \lambda) = \lambda(\lambda - 1)^{n-1}$.

Proof   Let G be a tree. We prove that $f(G; \lambda) = \lambda(\lambda - 1)^{n-1}$ by induction on n. If $n = 1$, the result is trivial. So assume the result for trees with at most $(n - 1)$ vertices, $n \geq 2$. Let G be a tree with n vertices, and e be a pendant edge of G. By Theorem 7.7.1, $f(G; \lambda) = f(G{-}e; \lambda) - f(G.e; \lambda)$. Now, $G{-}e$ is a forest with two component trees of orders $(n-1)$ and 1, and hence $f(G{-}e; \lambda) = (\lambda(\lambda-1)^{n-2})\lambda$ (see Exercise 7.1). Since $G.e$ is a tree with $(n - 1)$ vertices, $f(G.e; \lambda) = \lambda(\lambda - 1)^{n-2}$. Thus, $f(G; \lambda) = (\lambda(\lambda - 1)^{n-2})\lambda - \lambda(\lambda - 1)^{n-2} = \lambda(\lambda - 1)^{n-1}$.

Conversely, assume that G is a simple graph with $f(G; \lambda) = \lambda(\lambda - 1)^{n-1} = \lambda^n - (n - 1)\lambda^{n-1} + \cdots + (-1)^{n-1}\lambda$. Hence, by Theorem 7.7.2, G has n vertices and $(n - 1)$ edges. Further, the last term, $(-1)^{n-1}\lambda$, ensures that G is connected (see Exercise 7.2). Hence, G is a tree (see Theorem 4.1.4). $\square$

Remark 7.7.4   Theorem 7.7.3 shows that the chromatic polynomial of a graph G does not fix the graph uniquely up to isomorphism. For example, even though the graphs $K_{1,3}$ and P_4 are not isomorphic, they have the same chromatic polynomial, namely, $\lambda(\lambda - 1)^3$.

Exercise 7.2   If G has ω components, then show that λ^ω is a factor of $f(G; \lambda)$.

Exercise 7.3   Show that there exists no graph with the following polynomials as chromatic polynomials (i) $\lambda^5 - 4\lambda^4 + 8\lambda^3 - 4\lambda^2 + \lambda$; (ii) $\lambda^4 - 3\lambda^3 + \lambda^2$; (iii) $\lambda^7 - \lambda^6 + 1$.

Exercise 7.4   Find a graph G whose chromatic polynomial is $\lambda^5 - 6\lambda^4 + 11\lambda^3 - 6\lambda^2$.

Exercise 7.5   Show that for the cycle C_n of length n, $f(C_n; \lambda) = (\lambda - 1)^n + (-1)^n(\lambda - 1)$, $n \geq 3$.

Exercise 7.6   Show that for any graph G, $f(G \vee K_1; \lambda) = \lambda f(G; \lambda - 1)$ and hence prove that $f(W_n; \lambda) = \lambda(\lambda - 2)^n + (-1)^n\lambda(\lambda - 2)$.

Notes

A good reference for graph colorings is the book by Jensen and Toft [72]. The book by Fiorini and Wilson [39] concentrates on edge colorings. Theorem 7.3.3

(Myceilski's theorem) has also been proved independently by Blanche Descartes [31] as well as by Zykov [123].

The proof of Brooks's theorem given in this chapter is based on the proof given by Fournier [44] (see also references [19] and [68]).

The term "snark" was given to the snark graph by Martin Gardner after the unusual creature that is described in Lewis Carroll's poem, "The Hunting of the Snark." A detailed account of the snarks, including their constructions, can be found in the interesting book by Holton and Sheehan [68].

VIII
Planarity

8.0. Introduction

The study of planar and nonplanar graphs, and, in particular, the several attempts to solve the *four-color conjecture* have contributed a great deal to the growth of graph theory. Actually, these efforts had been instrumental to the development of algebraic, topological, and computational techniques in graph theory.

In this chapter, we present some of the basic results on planar graphs. In particular, the two important characterization theorems for planar graphs, namely, Wagner's theorem (same as the Harary-Tutte theorem) and Kuratowski's theorem are presented. Moreover, the nonhamiltonicity of the Tutte graph on 46 vertices (see Figure 8.28 as also the front wrapper) is also explained in detail.

8.1. Planar and Nonplanar Graphs

Definition 8.1.1 A graph G is *planar* if there exists a drawing of G in the plane in which no two edges intersect in a point other than a vertex of G, where each edge is a simple arc or a Jordan arc. Such a drawing of a planar graph G is called a *plane representation* of G. In this case, we also say that G has been embedded in the plane. A *plane graph* is a planar graph that has already been embedded in the plane.

Example 8.1.2 There exist planar as well as nonplanar graphs. In Figure 8.1, a planar graph and two of its plane representations are shown. Note that all trees

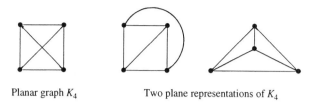

Planar graph K_4 Two plane representations of K_4

FIGURE 8.1. A planar graph with two plane representations

FIGURE 8.2. Arc connecting point x in int J with point y in ext J

are planar as also are cycles and wheels. The Petersen graph is nonplanar (a proof of this result will be given in the sequel).

Before proceeding further, let us recall here the celebrated Jordan curve theorem. If J is any closed Jordan curve in the plane, the complement of J (with respect to the plane) is partitioned into two disjoint open connected subsets of the plane, one of which is bounded and the other unbounded. The bounded subset is called the *interior* of J and is denoted by int J. The unbounded subset is called the *exterior* of J and is denoted by ext J. The *Jordan curve theorem* (of topology) states that any arc joining a point of int J and a point of ext J must intersect J at some point (see Figure 8.2) (the proof of this result, although intuitively obvious, is tedious).

Let G be a plane graph. Then the union of the edges (as Jordan arcs) of a cycle C of G form a closed Jordan curve, which we also denote by C. A plane graph G divides the rest of the plane (i.e., plane minus the edges and vertices of G), say π, into one or more faces, which we define below. We define an equivalence relation $\sim$ on π.

Definition 8.1.3   We say that for points A and B of π, $A \sim B$ if, and only if, there exists a Jordan arc from A to B in π. Clearly, $\sim$ is an equivalence relation on π. The equivalence classes of the above equivalence relation are called the *faces* of G.

Remarks 8.1.4

1. We claim that a connected graph is a tree if, and only if, it has only one face. Indeed, since there are no cycles in a tree T, the complement of a plane embedding of T in the plane is connected (in the above sense), and hence a tree has only one face. Conversely, it is clear that if a connected plane graph has only one face, then it must be a tree.

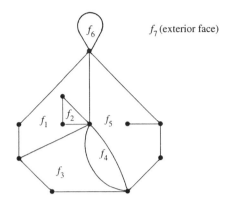

f_7 (exterior face)

FIGURE 8.3. A plane graph with seven faces

2. Any plane graph has exactly one unbounded face. The unbounded face is also referred to as the exterior face of the plane graph. All other faces, if any, are bounded. Figure 8.3 represents a plane graph with seven faces.

The distinction between bounded and unbounded faces of a plane graph is only superfluous, as there exists a plane representation of a plane graph G in which any specified face of G becomes the unbounded face. (This, of course, means that there exists a plane representation of G such that any specified vertex or edge belongs to the unbounded face.) To prove this result, we consider embeddings of a graph on a sphere. A graph is *embeddable on a sphere* S if it can be drawn on the surface of S so that its edges intersect only at its vertices. Such a drawing, if it exists, is called an embedding of G on S. Embeddings on a sphere are called *spherical embeddings*. What we have given here is only a naive definition. For a more rigorous description of spherical embeddings, see reference [55].

To prove the next theorem we need to recall the notion of stereographic projection. Let S be a sphere resting on a plane P. Let N be the "north pole," the point on the sphere diametrically opposite to the point of contact of S and P. Let the straight line joining N and a point s of $S\backslash\{N\}$ meet P at p. Then the mapping $\pi : S\backslash\{N\} \to P$ defined by $\pi(s) = p$ is called the *stereographic projection of S from N* (Figure 8.4).

Theorem 8.1.5 *A graph is planar if, and only if, it is embeddable on a sphere.*

Proof Let a graph G be embeddable on a sphere and let G' be a spherical embedding of G. The image of G' under the stereographic projection π of the sphere from a point N of the sphere not on G' is a plane representation of G on P. Conversely, if G' is a plane embedding of G on a plane P, then the inverse of the stereographic projection of G' on a sphere touching the plane P gives a spherical embedding of G. □

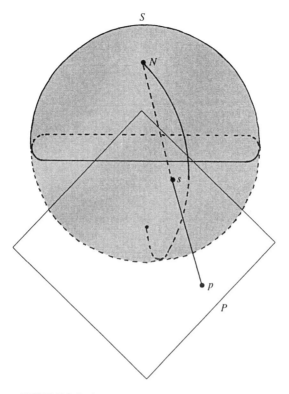

FIGURE 8.4. Stereographic projection of S from N

Theorem 8.1.6

a. *Let G be a plane graph and f be a face of G. Then there exists a plane embedding of G in which f is the exterior face.*

b. *Let G be a planar graph. Then G can be embedded in the plane in such a way that any specified vertex (or edge) belongs to the unbounded face of the resulting plane graph.*

Proof

a. Let n be a point of int f. Let G' be a spherical embedding σ of G, and let $N = \sigma(n)$. Let τ be the stereographic projection of the sphere with N as the north pole. Then the map $\tau\sigma$ (σ followed by τ) gives a plane embedding of G that maps f onto the exterior face of the plane representation $(\tau\sigma)(G)$ of G.

b. Let f be a face containing the specified vertex (respectively, edge) in a plane representation of G. Now, by part (a) of the theorem, there exists a plane embedding of G in which f becomes the exterior face. The specified vertex (respectively, edge) then becomes a vertex (respectively, edge) of the new unbounded face. □

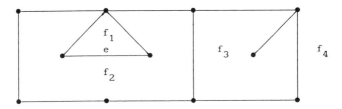

FIGURE 8.5. Plane graph G with four faces

Remarks 8.1.7

1. Let G be a connected plane graph. Each edge of G belongs to one or two faces of G. A cut edge of G belongs to exactly one face, and conversely, if an edge belongs to exactly one face of G, it must be a cut edge of G. An edge of G that is not a cut edge belongs to exactly two faces, and conversely.

2. The vertices and edges of a plane graph G belonging to a face of G are said to be *incident* with that face. The union of the vertices and edges of G incident with a face f of G is called the *boundary* of f and is denoted by $b(f)$. If G is connected, the boundary of each face is a closed walk in which each cut edge of G is traversed twice. When there are no cut edges, the boundary of each face of G is a closed trail in G. (See face f_1 of Figure 8.3.) However, if G is a disconnected plane graph, then the edges and the vertices incident with the exterior face will not define a trail.

3. The number of edges incident with a face f is defined as the *degree* of f. In counting the degree of a face, a cut edge is counted twice. Thus each edge of a plane graph G contributes 2 to the sum of the degrees of the faces. It follows that if $\mathcal{F}$ denotes the set of faces of a plane graph G, then $\sum_{f \in \mathcal{F}} d(f) = 2m(G)$, where $d(f)$ denotes the degree of the face f.

In Figure 8.5, $d(f_1) = 3$, $d(f_2) = 9$, $d(f_3) = 6$ and $d(f_4) = 8$.

Theorem 8.1.8 connects the planarity of G with the planarity of its blocks.

Theorem 8.1.8   *A graph G is planar if, and only if, each of its blocks is planar.*

Proof If G is planar, then each of its blocks is planar, since a subgraph of a planar graph is planar. Conversely, suppose that each block of G is planar. We now use induction on the number of blocks of G to prove the result. Without loss of generality, we assume that G is connected. If G has only one block, then G itself is a block, and hence G is planar.

Now, suppose that G has k planar blocks and that the result has been proved for all connected graphs having $(k - 1)$ planar blocks. Choose any end block B_0 of G and delete from G all the vertices of B_0 except the unique cut vertex, say v_0 of G in B_0. The resulting connected subgraph G' of G contains $(k - 1)$ planar blocks. Hence, by the induction hypothesis, G' is planar. Let $\tilde{G}'$ be a plane embedding of G' such that v_0 belongs to the boundary of the unbounded face, say f' (refer to

Theorem 8.1.6). Let $\tilde{B}_0$ be a plane embedding of B_0 in f' so that v_0 is in the exterior face of $\tilde{B}_0$. Then (by the identification of v_0 in the two embeddings), $\tilde{G}' \cup \tilde{B}_0$ is a plane embedding of G. □

Remark 8.1.9 In testing for the planarity of a graph G, one may delete multiple edges and loops of G, if any. This is so because, if a graph H is nonplanar, the removal of loops and parallel edges of H results in a subgraph of H, which is also nonplanar. Also, by Theorem 8.1.8, G can be assumed to be a block and hence 2-connected. If G has a vertex of degree 2, say v_0, and vv_0v' is the path formed by the two edges incident with v_0, contraction of vv_0 again results in a planar graph. Let G' be the graph obtained from G by performing such contractions successively at vertices of degree 2. Then G is planar if, and only if, G' is planar. From these observations, it is clear that in designing a planarity algorithm (i.e., an algorithm to test planarity), it suffices to consider only 2-connected simple graphs with minimum degree at least 3. (For a planarity algorithm, see reference [30]).

Exercise 1.1 Show that every graph with at most three cycles is planar.

Exercise 1.2 Find a simple graph G with degree sequence (4, 4, 3, 3, 3, 3) such that
(a) G is planar.
(b) G is nonplanar.

Exercise 1.3 Redraw the following planar graph so that the face f becomes the exterior face.

8.2. Euler Formula and Its Consequences

We have noted that a planar graph may have more than one plane representation (see Figure 8.1). A natural question that would arise is whether the number of faces is the same in each such representation. The answer to this question is provided by the Euler formula.

Theorem 8.2.1 (Euler formula)   *For a connected plane graph G, $n - m + f = 2$, where f denotes the number of faces of G.*

Proof   We apply induction on f.
If $f = 1$, then G is a tree and $m = n - 1$. Hence, $n - m + f = 2$.

Now, assume that the result is true for all plane graphs with $\mathfrak{f} - 1$ faces, $\mathfrak{f} \geq 2$, and suppose that G has $\mathfrak{f}$ faces. Since $\mathfrak{f} \geq 2$, G is not a tree, and hence contains a cycle C. Let e be an edge of C. Then e belongs to exactly two faces, say f_1 and f_2, of G and the deletion of e from G results in the formation of a single face from f_1 and f_2 (see Figure 8.5). Also, since e is not a cut edge of G, $G{-}e$ is connected. Further, the number of faces of $G{-}e$ is $\mathfrak{f} - 1$. So, by the induction hypothesis, $n - (m - 1) + (\mathfrak{f} - 1) = 2$, and this implies that $n - m + \mathfrak{f} = 2$. This completes the proof of the theorem. □

Below are some of the consequences of the Euler formula.

Corollary 8.2.2   *All plane embeddings of a given planar graph have the same number of faces.*

Proof   Since $\mathfrak{f} = m - n + 2$, the number of faces depends only on n and m, and not on the particular embedding.   □

Corollary 8.2.3   *If G is a simple planar graph with at least three vertices, then $m \leq 3n - 6$.*

Proof   Without loss of generality, we can assume that G is a simple connected plane graph. Since G is simple and $n \geq 3$, each face of G has degree at least 3. Hence, $\sum_{f \in \mathcal{F}} d(f) \geq 3\mathfrak{f}$. But $\sum_{f \in \mathcal{F}} d(f) = 2m$. Consequently, $2m \geq 3\mathfrak{f}$ that is, $\mathfrak{f} \leq \frac{2m}{3}$.

By the Euler formula, $m = n + \mathfrak{f} - 2$. Now, $\mathfrak{f} \leq \frac{2m}{3}$ implies that $m \leq n + (\frac{2m}{3}) - 2$. This gives that $m \leq 3n - 6$.   □

The above result is not valid if $n = 1$ or 2. Also, the condition of Corollary 8.2.3 is not sufficient for planarity of a simple connected graph, as the Petersen graph shows. For the Petersen graph, $m = 15$, $n = 10$, and hence $m \leq 3n - 6$, but the graph is not planar. (See Corollary 8.2.7 below.)

Example 8.2.4   Show that the complement of a simple planar graph with 11 vertices is nonplanar.

Solution   Let G be a simple planar graph with $n(G) = 11$. Since G is planar, $m(G) \leq 3n - 6 = 27$. If G^c were also planar, then, $m(G^c) \leq 3n - 6 = 27$. On the one hand, $m(G) + m(G^c) \leq 27 + 27 = 54$, whereas on the other hand, $m(G) + m(G^c) = m(K_{11}) = \binom{11}{2} = 55$. Hence, we arrive at a contradiction. This contradiction proves that G^c is nonplanar.

Corollary 8.2.5   *For any simple planar graph G, $\delta(G) \leq 5$.*

Proof   If $n \leq 6$, then $\Delta(G) \leq 5$. Hence, $\delta(G) \leq \Delta(G) \leq 5$, proving the result for such graphs. Hence, assume that $n \geq 7$. By Corollary 8.2.3, $m \leq 3n - 6$. Now,

$\delta n \leq \sum_{v \in V(G)} d_G(v) = 2m \leq 2(3n - 6) = 6n - 12$. Hence, $n(\delta - 6) \leq -12$. Consequently, $\delta - 6$ is negative, implying that $\delta \leq 5$.   □

Recall that the *girth* of a graph G is the length of a shortest cycle in G.

Theorem 8.2.6   *If the girth k of a connected plane graph G is at least 3, then* $m \leq \frac{k(n-2)}{(k-2)}$.

Proof   Let $\mathcal{F}$ denote the set of faces and f, as before, denote the number of faces of G. If $f \in \mathcal{F}$, then $d(f) \geq k$. Since $2m = \sum_{f \in \mathcal{F}} d(f)$, we get $2m \geq kf$. By Theorem 8.2.1, $f = 2 - n + m$. Hence, $2m \geq k(2 - n + m)$, implying that $m(k - 2) \leq k(n - 2)$. Thus, $m \leq \frac{k(n-2)}{(k-2)}$.   □

Corollary 8.2.7   *The Petersen graph P is nonplanar.*

Proof   The girth of the Petersen graph P is 5, $n(P) = 10$, and $m(P) = 15$. Hence, if P were planar, $15 \leq 5\frac{10-2}{5-2}$, which is not true. Hence, P is nonplanar.
   □

Exercise 2.1   Show that every simple bipartite cubic planar graph contains a C_4.

Exercise 2.2   A nonplanar graph G is called *planar vertex critical* if $G-v$ is planar for every vertex v of G. Prove that a planar vertex critical graph must be 2-connected.

Exercise 2.3   Verify Euler formula for the following plane graph.

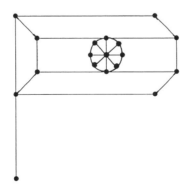

Exercise 2.4   Let G be a simple plane cubic graph having 8 faces. Determine $n(G)$. Draw two such graphs that are nonisomorphic.

Exercise 2.5   Prove that if G is a simple connected planar bipartite graph, then $m \leq 2n - 4$, where $n \geq 3$.

Exercise 2.6   Prove that a simple planar graph has at least 4 vertices of degree 5 at most.

Exercise 2.7   If G is a nonplanar graph, show that it has either 5 vertices of degree at least 4, or 6 vertices of degree at least 3.

Exercise 2.8   Prove that a planar graph with minimum degree at least 5 contains at least 12 vertices. Give an example of a planar graph on 12 vertices with minimum degree 5.

Exercise 2.9   Show that there is no 6-connected planar graph.

Exercise 2.10   Let G be a plane graph of order n and size m in which every face is bounded by a k-cycle. Show that $m = \frac{k(n-1)}{(k-2)}$.

Definition 8.2.8   A graph G is *maximal planar* if G is planar, but for any pair of nonadjacent vertices u and v of G, $G + (uv)$ is nonplanar.

Remark 8.2.9   *Any planar graph is a spanning subgraph of a maximal planar graph.* Indeed, if $\tilde{G}$ is a plane embedding of a planar graph G with at least three vertices, and if $e = uv$ is a cut edge of $\tilde{G}$ embedded in a face f of $\tilde{G}$, it is clear that there exists a vertex w on the boundary of f such that the edge uw or vw can be drawn in f so that either $\tilde{G} + (vw)$ or $\tilde{G} + (uw)$ is also a plane graph (see Figure 8.6(a)). Further, if C_0 is any cycle bounding a face f_0 of a plane graph H, then edges can be drawn in int C_0 without crossing each other so that f_0 is divided into triangles (see Figure 8.6(b)).

Definition 8.2.10   A *plane triangulation* is a plane graph in which each of its faces is bounded by a triangle. A plane triangulation of a plane graph G is a plane triangulation H such that G is a spanning subgraph of H.

Remark 8.2.11   Remark 8.2.9 shows that a plane embedding of a simple maximal planar graph is a plane triangulation.

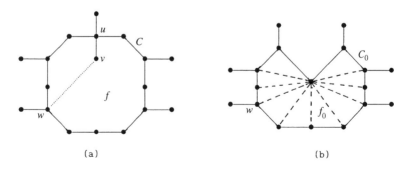

(a)   (b)

FIGURE 8.6. Procedure to get maximal planar graphs

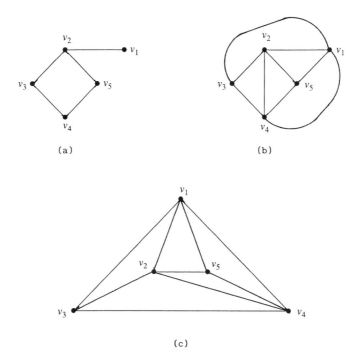

FIGURE 8.7. (a) Graph G and (b, c) plane triangulations of G

Note that any simple plane graph is a subgraph of a simple maximal plane graph and hence it is a spanning subgraph of some plane triangulation. Thus, to any simple plane graph G that is not already a plane triangulation, we can add a set of new edges to obtain a plane triangulation. The set of new edges thus added need not be unique.

Figure 8.7(a) is a simple plane graph G and Figure 8.7(b) is a plane triangulation of G; Figure 8.7(c) is a plane triangulation of G isomorphic to the graph of Figure 8.7(b) having only straight-line edges. (A result of Fáry [38] states that every simple planar graph has a plane embedding in which each edge is a straight line.)

Exercise 2.11   Embed the 3-cube Q_3 (see Exercise 3.4 of Chapter 5) in a maximal planar graph having the same vertex set as Q_3. Count the number of new edges added.

Exercise 2.12   Prove that for a simple maximal planar graph on $n \geq 3$ vertices, $m = 3n - 6$.

Exercise 2.13   Use Exercise 2.12 to show that for any simple planar graph, $m \leq 3n - 6$.

Exercise 2.14   Show that every plane triangulation of order $n \geq 4$ is 3-connected.

Exercise 2.15 Let G be a maximal planar graph with $n \geq 4$. Let n_i denote the number of vertices of degree i in G. Then prove that $3n_3 + 2n_4 + n_5 = 12 + n_7 + 2n_8 + 3n_9 + 4n_{10} + \cdots$ (Hint: Use the fact that $n = n_3 + n_4 + n_5 + n_6 + \cdots$).

Exercise 2.16   Generalize the Euler formula for disconnected plane graphs.

8.3. K_5 and $K_{3,3}$ are Nonplanar Graphs

In this section we prove that K_5 and $K_{3,3}$ are nonplanar. These two graphs are basic in Kuratowski's characterization of planar graphs (see Theorem 8.6.5 later in this chapter). For this reason, they are often referred to as the two *Kuratowski graphs*.

Theorem 8.3.1 K_5 *is nonplanar.*

First Proof   This proof uses the Jordan curve theorem. Assume the contrary, namely, K_5 is planar. Let v_1, v_2, v_3, v_4, and v_5 be the vertices of K_5 in a plane representation of K_5. The cycle $C = v_1 v_2 v_3 v_4 v_1$ (as a closed Jordan curve) divides the plane into two faces, namely, the interior and the exterior of C. The vertex v_5 must belong either to int C or to ext C. Suppose that v_5 belongs to int C (a similar proof holds if v_5 belongs to ext C). Draw the edges $v_5 v_1$, $v_5 v_2$, $v_5 v_3$, and $v_5 v_4$ in int C. Now there remain two more edges $v_1 v_3$ and $v_2 v_4$ to be drawn. None of these can be drawn in int C, since it is assumed that K_5 is planar. Thus, $v_1 v_3$ lies in ext C. Then one of v_2 and v_4 belongs to the interior of the cycle $C_1 = v_1 v_5 v_3 v_1$ and the other to its exterior (see Figure 8.8). Hence, $v_2 v_4$ cannot be drawn without violating planarity.   □

Remark 8.3.2   The first proof of Theorem 8.3.1 shows that all the edges of K_5 except one can be drawn in the plane without violating planarity. Hence, for any edge e of K_5, $K_5 - e$ is planar.

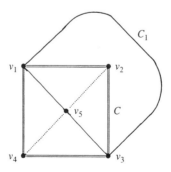

FIGURE 8.8. Graph for first proof of Theorem 8.3.1

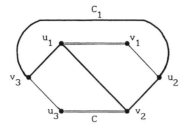

FIGURE 8.9. Graph for first proof of Theorem 8.3.3

Second Proof If K_5 were planar, it follows from Theorem 8.2.6 that $10 \leq \frac{3(5-2)}{(3-2)} = 9$, which is not true. Hence, K_5 is nonplanar. □

Theorem 8.3.3 $K_{3,3}$ *is nonplanar.*

First Proof Proof is by the use of the Jordan curve theorem. Suppose that $K_{3,3}$ is planar. Let $U = \{u_1, u_2, u_3\}$ and $V = \{v_1, v_2, v_3\}$ be the bipartition of $K_{3,3}$ in a plane representation of the graph. Consider the cycle $C = u_1 v_1 u_2 v_2 u_3 v_3 u_1$. Since the graph is assumed to be planar, the edge $u_1 v_2$ must lie either in the interior of C or in its exterior. For the sake of definiteness, assume that it lies in int C (a similar proof holds if one assumes that the edge $u_1 v_2$ lies in ext C). Two more edges remain to be drawn, namely, $u_2 v_3$ and $u_3 v_1$. None of these can be drawn in int C without crossing the edge $u_1 v_2$. Hence, both of them are to be drawn in ext C. Now draw $u_2 v_3$ in ext C. Then one of v_1 and u_3 belongs to the interior of $C_1 = u_1 v_2 u_2 v_3 u_1$ and the other to the exterior of C_1 (see Figure 8.9). Hence, the edge $v_1 u_3$ cannot be drawn without violating planarity. This shows that $K_{3,3}$ is nonplanar. □

Second Proof If $K_{3,3}$ were planar, let f be the number of faces of $G = K_{3,3}$ in a plane embedding of G and $\mathcal{F}$, the set of faces of G. As the girth of $K_{3,3}$ is 4, we have $m = \frac{1}{2} \sum_{f \in \mathcal{F}} d(f) \geq \frac{4f}{2} = 2f$. By Theorem 8.2.1, $n - m + f = 2$. For $K_{3,3}$, $n = 6$, and $m = 9$. Hence, $f = 2 + m - n = 5$. Thus, $9 \geq 2.5 = 10$, a contradiction. □

Exercise 3.1 Give yet another proof of Theorem 8.3.3.

Exercise 3.2 Find the maximum number of edges in a planar complete tripartite graph with each part size at least 2.

Remark 8.3.4 As in the case of K_5, for any edge e, $K_{3,3}-e$ is planar. Observe that the graphs K_5 and $K_{3,3}$ have some features in common.

1. Both are regular graphs.
2. Removal of a vertex or an edge from each graph results in a planar graph.
3. Contraction of an edge results in a planar graph.

4. K_5 is a nonplanar graph with the smallest number of vertices, whereas $K_{3,3}$ is a nonplanar graph with the smallest number of edges. (Hence, any nonplanar graph must have at least 5 vertices and 9 edges.)

8.4. Dual of a Plane Graph

Let G be a plane graph. One can form a new graph H in the following way. Corresponding to each face f of G, take a vertex f^* and an edge e^* corresponding to each edge e of G. Then edge e^* joins vertices f^* and g^* in H if, and only if, edge e is common to the boundaries of faces f and g in G. (It is possible that f may be the same as g.) Graph H is then called the *dual* (or more precisely, the geometric dual) of G. (Figure 8.10). If e is a cut edge of G embedded in face f of G, then e^*

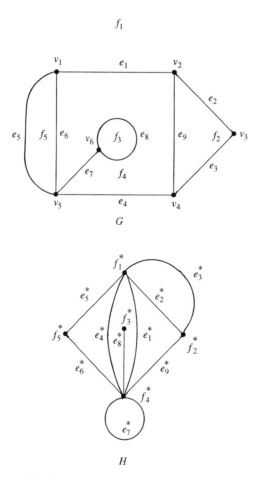

FIGURE 8.10. A plane graph G and its dual H

is a loop at f^*. H is a planar graph, and there exists a natural way of embedding H in the plane. Vertex f^*, corresponding to face f, is placed in face f of G. Edge e^*, joining f^* and g^*, is drawn so that e^* crosses e once and only once and crosses no other edge. This procedure is illustrated in Figure 8.11. This embedding is the canonical embedding of H. H with this canonical embedding is denoted by G^*. Any two embeddings of H, as described above, are isomorphic.

The definition of the dual implies that $m(G^*) = m(G), n(G^*) = f(G)$, and $d_{G^*}(f^*) = d_G(f)$, where $d_G(f)$ denotes the degree of face f of G.

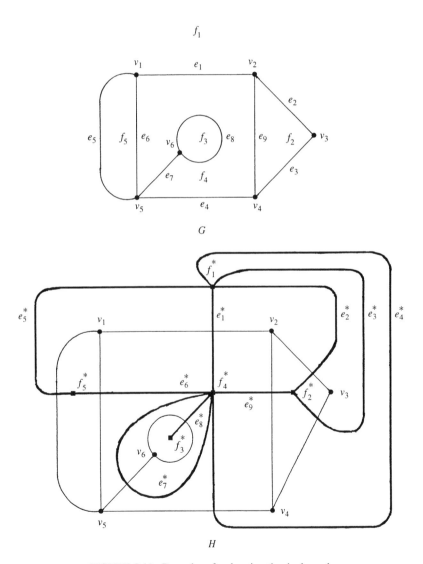

FIGURE 8.11. Procedure for drawing the dual graph

From the manner of construction of G^*, it follows that

(i)   an edge e of a plane graph G is a cut edge of G if, and only if, e^* is a loop
      of G^*, and it is a loop if, and only if, e^* is a cut edge of G^*, and
(ii) G^* is connected whether G is connected or not (see graphs G and G^* of
      Figure 8.12).

The canonical embedding of the dual of G^* is denoted by G^{**}. It is easy to check
that G^{**} is isomorphic to G, if, and only if, G is connected. Graph isomorphism

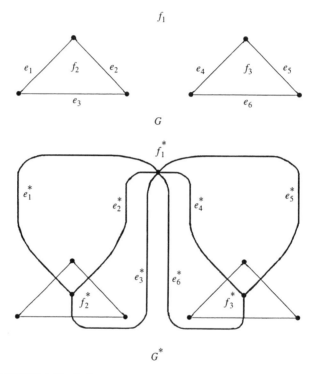

FIGURE 8.12. A disconnected graph G and its (connected) dual G^*

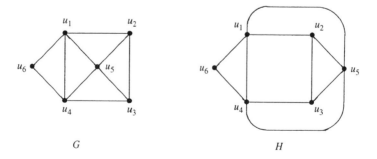

FIGURE 8.13. Isomorphic graphs G and H for which $G^* \not\cong H^*$

does not preserve duality; that is, isomorphic plane graphs may have nonisomorphic duals. The graphs G and H of Figure 8.13 are isomorphic plane graphs but $G^* \not\cong H^*$. G has a face of degree 5, whereas no face of H has degree 5. Hence, G^* has a vertex of degree 5, whereas H^* has no vertex of degree 5. Consequently, $G^* \not\cong H^*$.

Exercise 4.1   Draw the dual of
(i)   the Herschel graph (graph of Figure 5.4)
(ii)   the graph G given below:

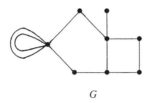

G

Exercise 4.2   A plane graph G is called *self-dual* if $G \cong G^*$. Prove the following:
(i)   All wheels $W_n (n \geq 3)$ are self-dual.
(ii)   For a self-dual graph, $2n = m + 2$.

Exercise 4.3   Construct two infinite families of self-dual graphs.

8.5.   The Four-Color Theorem and the Heawood Five-Color Theorem

What is the minimum number of colors required to color the world map of countries so that no two countries having a common boundary receive the same color? This simple-looking problem manifested itself into one of the most challenging problems of graph theory, popularly known as the four-color conjecture (4CC).

The geographical map of the countries of the world is a typical example of a plane graph. An assignment of colors to the faces of a plane graph G so that no two faces having a common boundary containing at least one edge receive the same color is a *face coloring* of G. The face chromatic number $\chi^*(G)$ of a plane graph G is the minimum k for which G has a face coloring using k colors. The problem of coloring a map so that no two adjacent countries receive the same color can thus be transformed into a problem of face coloring of a plane graph G. The face coloring of G is closely related to the vertex coloring of G^*. The fact that any two faces of G are adjacent in G if, and only if, the corresponding vertices of G^* are adjacent in G^* shows that G is k-face colorable if, and only if, G^* is k-vertex colorable.

It was young Francis Guthrie who conjectured, while coloring the district map of England, that four colors were sufficient to color the world map so that adjacent countries receive distinct colors. This conjecture was communicated by his brother to De Morgan in 1852. The conjecture of Guthrie is equivalent to the conjecture that any plane graph is 4-face colorable. The latter conjecture is equivalent to the conjecture: Every planar graph is 4-vertex colorable.

Ever since the conjecture was first published in 1852, many eminent mathematicians, especially graph theorists, attempted to settle the conjecture. In the process of settling the conjecture, many equivalent formulations of this conjecture were found. Assaults on the conjecture were made using such varied branches of mathematics as algebra, number theory, and finite geometries. The solution found the light of the day when K. Appel, W. Haken, and J. Koch [4] of the University of Illinois established the validity of the conjecture in 1976 with the aid of computers (see also references [2] and [3]). The proof includes, among other things, 10^{10} units of operations, amounting to a staggering 1200 hours of computer time on the high-speed computer available at that time.

Although the computer-oriented proof of Appel, Haken, and Koch settled the conjecture in 1976 and has stood the test of time, a theoretical proof of the four-color problem is still to be found.

Even though the solution of the 4CC has been a formidable task, it is rather easy to establish that every planar graph is 6-vertex colorable.

Theorem 8.5.1   *Every planar graph is 6-vertex colorable.*

Proof   The proof is by induction on n, the number of vertices of the graph. The result is trivial for planar graphs with at most 6 vertices. Assume the result for planar graphs with $n-1$, $n \geq 7$, vertices. Let G be a planar graph with n vertices. By Corollary 8.2.5, $\delta(G) \leq 5$, and thus G has a vertex v of degree at most 5. By hypothesis, $G-v$ is 6-vertex colorable. In any 6-vertex coloring of $G-v$, the neighbors of v in G would have used only at most five colors, and hence v can be colored by an unused color. In other words, G is 6-vertex colorable.   □

It involves some ingenious arguments to reduce the upper bound for the chromatic number of a planar graph from 6 to 5. The upper bound 5 was obtained by Heawood [67] as early as 1890.

Theorem 8.5.2 (Heawood five-color theorem)   *Every planar graph is 5-vertex colorable.*

Proof   Proof is by induction on $n(G) = n$. Without loss of generality, we assume that G is a connected plane graph. If $n \leq 5$, the result is clearly true. Hence, assume that $n \geq 6$ and that any planar graph with fewer than n vertices is 5-vertex colorable. G being planar, $\delta(G) \leq 5$ by Corollary 8.2.5, and so G contains a vertex v_0 of degree not exceeding 5. By induction hypothesis, $G-v_0$ is 5-vertex colorable.

If $d(v_0) \leq 4$, at most four colors would have been used in coloring the neighbors of v_0 in G in a 5-vertex coloring of $G-v_0$. Hence, an unused color can then be assigned to v_0 to yield a proper 5-vertex coloring of G.

If $d(v_0) = 5$, but only four or fewer colors are used to color the neighbors of v_0 in a proper 5-vertex coloring of $G-v_0$, then also an unused color can be assigned to v_0 to yield a proper 5-vertex coloring of G.

Hence, assume that the degree of v_0 is 5 and that in every 5-coloring of $G-v_0$, the neighbors of v_0 in G receive five distinct colors. Let v_1, v_2, v_3, v_4, and v_5 be the neighbors of v_0 in a cyclic order in a plane embedding of G. Choose some proper 5-coloring of $G-v_0$ with colors, say $c_1, c_2, \ldots, c_5$. Let $\{V_1, V_2, \ldots, V_5\}$ be the color partition of $G-v_0$ where the vertices in V_i are colored $c_i, 1 \leq i \leq 5$. Assume further that $v_i \in V_i, 1 \leq i \leq 5$.

Let G_{ij} be the subgraph of $G-v_0$ induced by $V_i \cup V_j$. Suppose v_i and $v_j, 1 \leq i, j \leq 5$, belong to distinct components of G_{ij}. Then the interchange of the colors c_i and c_j in the component of G_{ij} containing v_i would give a recoloring of $G-v_0$ in which only four colors are assigned to the neighbors of v_0. But this is against our assumption. Hence, v_i and v_j must belong to the same component of G_{ij}. Let P_{ij} be a v_i-v_j path in G_{ij}. Let C denote the cycle $v_0 v_1 P_{13} v_3 v_0$ in G (Figure 8.14). Then C separates v_2 and v_4; that is, one of v_2 and v_4 must lie in int C and the other in ext C. In Figure 8.14, $v_2 \in$ int C and $v_4 \in$ ext C. Then P_{24} must cross C at a vertex of C. But this is clearly impossible since no vertex of C receives either of the colors c_2 and c_4. Hence, this possibility cannot arise and, G is 5-vertex colorable.   □

Note that the bound 4 in the inequality $\chi(G) \leq 4$ for planar graphs G is best possible since K_4 is planar and $\chi(K_4) = 4$.

Exercise 5.1   Show that a planar graph G is bipartite if, and only if, each of its faces is of even degree in any plane embedding of G.

Exercise 5.2   Show that a connected plane graph G is bipartite if, and only if, G^* is Eulerian. Hence show that a connected plane graph is 2-face colorable if, and only if, it is Eulerian.

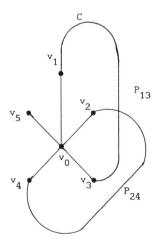

FIGURE 8.14. Graph for proof of Theorem 8.5.2

Exercise 5.3   Prove that a Hamiltonian plane graph is 4-face colorable and that its dual is 4-vertex colorable.

Exercise 5.4   Show that a plane triangulation has a 3-face coloring if, and only if, it is not K_4. (Hint: Use Brooks's theorem).

Remark 8.5.3   (Grötzsch):   If G is a planar graph that contains no triangle, then G is 3-vertex colorable.

8.6.   Kuratowski's Theorem

Definitions 8.6.1

1. A *subdivision of an edge* $e = uv$ of a graph G is obtained by introducing a new vertex w in e, that is, by replacing the edge $e = uv$ of G by the path uwv of length 2 so that the new vertex w is of degree 2 in the resulting graph (see Figure 8.15(a)).

2. A *homeomorph* or a *subdivision of a graph* G is a graph obtained from G by applying a finite number of subdivisions of edges in succession (see Figure 8.15(b)). G itself is regarded as a subdivision of G.

3. Two graphs G_1 and G_2 are called *homeomorphic* if they are both homeomorphs of some graph G. Clearly, the graphs of Figure 8.15(b) are homeomorphic, even though neither of the two graphs is a homeomorph of the other.

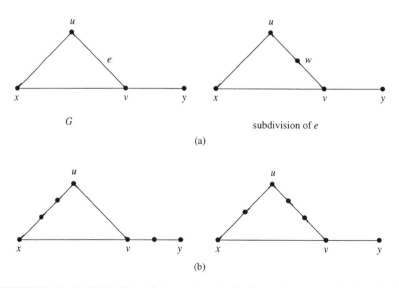

FIGURE 8.15. (a) Subdivision of edge e of graph G, (b) two homeomorphs of graph G

Kuratowski's theorem [81] characterizing planar graphs was one of the major breakthrough results in graph theory of the 20th century. As mentioned earlier, while examining planarity of graphs, we need only consider simple graphs since the presence of loops and multiple edges does not affect the planarity of graphs. Consequently, *a graph is planar if, and only if, its underlying simple graph is planar.* We therefore consider in this section only (finite) simple graphs. We recall that for any edge e of a graph G, $G-e$ is the subgraph of G obtained by deleting the edge e, whereas $G \cdot e$ denotes the contraction of e. We always discard isolated vertices when edges get deleted and remove the new multiple edges when edges get contracted. More generally, for a subgraph H of G, $G \cdot H$ denotes the graph obtained by the successive contractions of all the edges of H in G. The resulting graph is independent of the order of contraction. Moreover, if G is planar, then $G \cdot e$ is planar; consequently, $G \cdot H$ is planar. In other words, if $G \cdot H$ is nonplanar for some subgraph H of G, then G also is nonplanar. Moreover, any two homeomorphic graphs are contractible to the same graph.

Definition 8.6.2 If $G \circ H = K$, we call K a *contraction* of G; we also say that G is *contractible* to K. G is said to be *subcontractible* to K if G has a subgraph H contractible to K. We also refer to this fact by saying that K is a *subcontraction* of G.

Example 8.6.3 For instance, in Figure 8.16, graph G is subcontractible to triangle abc. (Take H to be the cycle $abcd$ and contract the edge ad in H. By abuse of notation, the new vertex is denoted by a or d.) We note further that if G' is a homeomorph of G, then contraction of one of the edges incident at each vertex of degree 2 in $V(G') \backslash V(G)$ results in a graph homeomorphic to G.

Our first aim is to prove the following result, which was established by Wagner [119] and, independently, by Harary and Tutte [64].

Theorem 8.6.4 *A graph is planar if, and only if, it is not subcontractible to K_5 or $K_{3,3}$.*

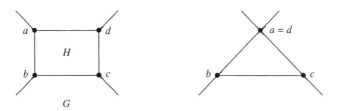

FIGURE 8.16. Graph G subcontractible to triangle abc

As a consequence, we establish Kuratowski's characterization theorem for planar graphs.

Theorem 8.6.5 (Kuratowski [81])   *A graph is planar if, and only if, it has no subgraph homeomorphic to K_5 or $K_{3,3}$.*

The proofs of Theorems 8.6.4 and 8.6.5, as presented here, are due to Fournier [45]. Recall that any subgraph and any contraction of a planar graph are both planar.

Definition 8.6.6   Let us call a simple connected nonplanar graph G *irreducible* if, for each edge e of G, $G \cdot e$ is planar.

For instance, both K_5 and $K_{3,3}$ are irreducible.

Proof of Theorem 8.6.4   If G has a subgraph G_0 contractible to K_5 or $K_{3,3}$, then, since K_5 and $K_{3,3}$ are nonplanar, G_0 and therefore G are nonplanar.

We now prove the converse. Assume that G is a simple connected nonplanar graph. By Theorem 8.1.8, at least one block of G is nonplanar. Hence, assume that G is a simple 2-connected nonplanar graph. We now show that G has a subgraph contractible to K_5 or $K_{3,3}$.

Keep contracting edges of G (and delete the new multiple edges, if any, at each stage of the contraction) until there results a (2-connected) irreducible (nonplanar) graph H. Clearly, $\delta(H) \geq 3$. Now, if e and f are any two distinct edges of G, then $(G \cdot e) - f = (G - f) \cdot e$. Hence, the graph H may as well be obtained by deleting a set (which may be empty) of edges of G, resulting in a subgraph G_0 of G and then contracting a subgraph of G_0. We now complete the proof of the theorem by showing that H has a subgraph K homeomorphic (and hence contractible) to K_5 or $K_{3,3}$. In this case, G has the subgraph G_0, which is contractible to K_5 or $K_{3,3}$.

Let $e = ab \in E(H)$ and $H' = H - \{a, b\}$. Then H' is connected; otherwise, let $G_1', G_2', \ldots, G_t' (t \geq 2)$ be the components of H' (see Figure 8.17). Let $H_{t-1}' = H \cdot G_{t-1}'$ and $H_t' = H \cdot G_t'$. Since H is irreducible, and $\delta(H) \geq 3$, H_{t-1}' and H_t' are planar. Let w_t' be the vertex to which G_t' gets contracted. Then w_t' is a vertex of degree 2 in H_t'. Let π_t and π_{t-1} be plane embeddings of H_t' and H_{t-1}', respectively. In π_t, we can assume without loss of generality that the edges aw_t' and $w_t'b$ belong to the boundary of the exterior face (see Theorem 8.1.6). We now replace the path $aw_t'b$ in π_t by the plane embedding of $G_t' \cup [V(G_t'), \{a, b\}]$ contained in the embedding π_{t-1}. (Recall the notation that $[S_1, S_2]$ denotes the set of edges each having one end in S_1 and the other end in S_2.) This gives a plane embedding of H, a contradiction. Thus H' is connected.

Case 1. H' has a cut vertex v. Let $G_1, G_2, \ldots, G_r (r \geq 2)$ be the components of $H' - \{v\}$, and let $G_1, G_2, \ldots, G_s, 0 \leq s \leq r$, be those components that are connected to both a and b in H (see Figure 8.18). Without loss of generality, assume that G_r is connected to b. If $r > s$, then G_r must have at least one edge. If not, G_r will consist of a single vertex, say z. Then z will be adjacent only to v and

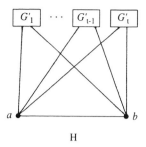

FIGURE 8.17. Graph for proof of Theorem 8.6.4

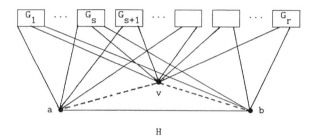

FIGURE 8.18. Graph of H for case 1 of proof of Theorem 8.6.4

b and hence z is a vertex of degree 2 in H. But this is a contradiction to the fact that $\delta(H) \geq 3$. Let $H_r = H \cdot G_r$ and $H_{r-1} = H \cdot e_{r-1}$, where e_{r-1} is an edge with one end v and the other end in $V(G_{r-1})$. Since H is irreducible, both H_r and H_{r-1} are planar. Let w_r be the vertex to which G_r gets contracted. Then w_r is a vertex of degree 2 in H_r. Let π_r and π_{r-1} be plane embeddings of H_r and H_{r-1}, respectively. In π_r, we can assume without loss of generality that the edges vw_r and w_rb belong to the boundary of the exterior face. We now replace the path vw_rb in π_r by the plane embedding of $G_r \cup [V(G_r), \{v, b\}]$ contained in the embedding π_{r-1}. This gives a plane embedding of H, a contradiction. Thus, $r = s$. If $r = s = 2$, any plane embedding of $H \cdot G_1$ and any plane embedding of $H \cdot G_2$ yield a plane embedding of H (see Figure 8.19). This shows that $r = s \geq 3$. In this case, H contains a homeomorph of $K_{3,3}$ (see Figure 8.20) with $\{w_1, w_2, w_3; a, b, v\}$ being the vertex set of $K_{3,3}$. (It is clear that since the G_i's are connected, the vertices w_1, w_2, and w_3 shown in Figure 8.20 must exist.)

   Case 2. H' is 2-connected. H' contains a cycle C of length at least 3. Consider a plane embedding of $H \cdot e$ (where $e = ab$, as above). If c denotes the new vertex to which a and b get contracted, $(H \cdot e)-c = H'$. We may therefore suppose without loss of generality that c is in the interior of the cycle C in the plane embedding of $H \cdot e$.

   Now, the edges of $H \cdot e$ incident to c arise out of edges of H incident to a or b. There arise three possibilities with reference to the positions of the edges of $H \cdot e$ incident to c relative to the cycle C.

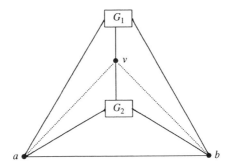

FIGURE 8.19. Plane embedding for case 1 of proof of Theorem 8.6.4

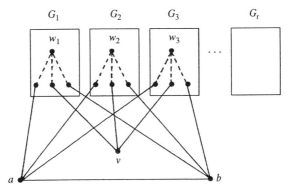

FIGURE 8.20. Homeomorph for case 1 of proof of Theorem 8.6.4

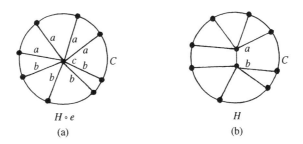

FIGURE 8.21. First configuration for case 2 of proof of Theorem 8.6.4. Edges incident to a and b are marked a and b, respectively

(i) Suppose the edges incident to c occur so that the edges incident to a and the edges incident to b in H are consecutive around c in a plane embedding of $H \cdot e$, as shown in Figure 8.21(a). Since H is a minimal nonplanar graph, the paths from c to C can be only single edges.) Then the plane representation of $H \cdot e$ gives a plane representation of H, as in Figure 8.21(b), a contradiction. So this possibility cannot arise.

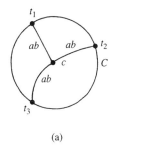
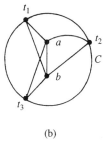

(a)

(b)

FIGURE 8.22. Second configuration for case 2 of proof of Theorem 8.6.4. Edges incident to both a and b are marked ab

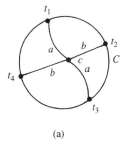
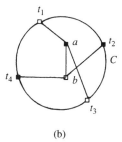

(a)

(b)

FIGURE 8.23. Third configuration for case 2 of proof of Theorem 8.6.4

(ii) Suppose there are three edges of $H \cdot e$ incident with c with each edge corresponding to a pair of edges of H, one incident to a and the other to b as in Figure 8.22(a). Then H contains a subgraph contractible to K_5, as shown in Fig. 8.22(b).

We are now left with only one more possibility.

(iii) There are four edges of $H \cdot e$ incident to c and they arise alternately out of edges incident to a and b in H, as in Figure 8.23(a). Then there arises in H a homeomorph of $K_{3,3}$, as shown in Figure 8.23(b). The sets $X = \{a, t_2, t_4\}$ and $Y = \{b, t_1, t_3\}$ are the sets of the bipartition of this homeomorph of $K_{3,3}$. □

We now proceed to prove Theorem 8.6.5.

Proof of Theorem 8.6.5 The "sufficiency" part of the proof is trivial. If G contains a homeomorph of either K_5 or $K_{3,3}$, G is certainly nonplanar, since a homeomorph of a planar graph is planar.

Assume that G is nonplanar. Remove edges from G one after another until we get an edge-minimal nonplanar subgraph G_0 of G; that is, G_0 is nonplanar and for any edge e of G, G_0-e is planar. Now contract the edges in G_0 incident with vertices of degree at most 2 in some order. Let us denote the resulting graph by G_0'. Then G_0' is nonplanar whereas $G_0'-e$ is planar for any edge e of G_0', and the minimum degree of G_0' is at least 3. We now have to show that G_0' contains a homeomorph of K_5 or $K_{3,3}$.

By Theorem 8.6.4, G_0' is subcontractible to K_5 or $K_{3,3}$. This means that G_0' contains a subgraph H, which is contractible to K_5 or $K_{3,3}$. As $G_0'-e$ is planar for any edge e of G_0', $G_0' = H$. Thus, G_0' itself is contractible to K_5 or $K_{3,3}$. If G_0' is either K_5 or $K_{3,3}$, we are done. Assume now that G_0' is neither K_5 nor $K_{3,3}$. Let $e_1, e_2, \ldots, e_r$ be the edges of G_0' contracted in order, resulting in a K_5 or $K_{3,3}$.

First, let us assume that $r = 1$ so that $G_0' \cdot e_1$ is either K_5 or $K_{3,3}$. Suppose that $G_0' \cdot e_1 = K_{3,3}$, with $\{x_1, x_2, x_3\}$ and $\{y_1, y_2, y_3\}$ as the partite sets of vertices. Suppose that x_1 is the vertex obtained by identifying the ends of e_1. We may then take $e_1 = x_1 a$ (by abuse of notation), where a is a vertex distinct from the x_i's and y_j's (Figure 8.24(a)). If a is adjacent to all of y_1, y_2, and y_3, then $\{a, x_2, x_3\}$ and $\{y_1, y_2, y_3\}$ form a bipartition of a $K_{3,3}$ in G_0'. If a is adjacent to only one or two of $\{y_1, y_2, y_3\}$ (Figure 8.24(b)), then again G_0' contains a homeomorph of $K_{3,3}$.

Next, let us assume that $G_0' \cdot e_1 = K_5$ with vertex set $\{v_1, v_2, v_3, v_4, v_5\}$. Suppose that v_1 is the vertex obtained by identifying the ends of e_1. As before, we may take $e_1 = v_1 a$, where $a \notin \{v_1, v_2, v_3, v_4, v_5\}$. If a is adjacent to all of $\{v_2, v_3, v_4, v_5\}$, then $G_0'-v_1$ is a K_5, a contradiction to the fact that any proper subgraph of G_0' is planar. If a is adjacent to only three of $\{v_2, v_3, v_4, v_5\}$,

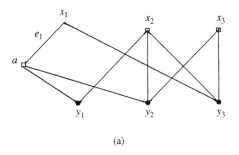

(a)

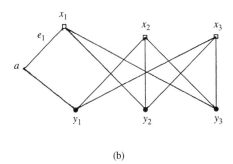

(b)

FIGURE 8.24. Graph for proof of Theorem 8.6.5

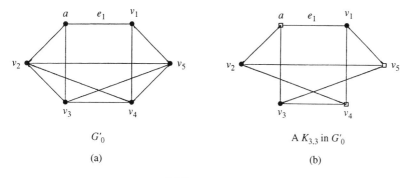

G'₀

(a)

A K₃,₃ in G'₀

(b)

FIGURE 8.25.

say v_2, v_3, and v_4, then the edge-induced subgraph of G'_0 induced by the edges $av_1, av_2, av_3, av_4, v_1v_5, v_2v_3, v_2v_4, v_2v_5, v_3v_4, v_3v_5,$ and v_4v_5 is a homeomorph of K_5. In this case, G'_0 also contains a homeomorph of $K_{3,3}$ since $d_{G'_0}(v_1) \geq 3$, v_1 is adjacent to at least one of v_2, v_3, and v_4, say v_2. Then the edge-induced subgraph of G'_0 induced by the edges in $\{av_1, av_3, av_4, v_1v_2, v_2v_3, v_2v_4, v_1v_5, v_3v_5, v_4v_5\}$ is a $K_{3,3}$. We now consider the case when a is adjacent to only two of v_2, v_3, v_4, and v_5, say v_2 and v_3. Then, necessarily, v_1 is adjacent to v_4 and v_5 (since on contraction of the edge v_1a, v_1 is adjacent to v_2, v_3, v_4, and v_5). In this case G'_0 also contains a $K_{3,3}$ (see Figure 8.25). Finally, the case when a is adjacent to at most one of v_2, v_3, v_4, and v_5 cannot arise since the degree of a is at least 3 in G'_0. Thus, in any case, we have proved that when $r = 1$, G'_0 contains a homeomorph of $K_{3,3}$.

The result can now easily be seen to be true by induction on r. Indeed, if $H_2 = H_1 \cdot e$ and H_2 contains a homeomorph of $K_{3,3}$, then H_1 contains a homeomorph of $K_{3,3}$. □

The nonplanarity of the Petersen graph (Figure 8.26(a)) can be established by showing that it is contractible to K_5 (see Figure 8.26(b)) or by showing that it contains a homeomorph of $K_{3,3}$ (see Figure 8.26(c)).

Exercise 6.1 Prove that the following graph is nonplanar.

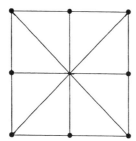

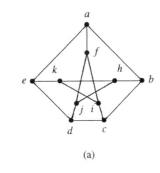

(a)

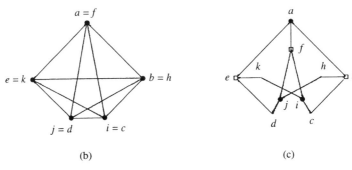

(b)   (c)

FIGURE 8.26. Nonplanarity of the Petersen graph. (a) The Petersen graph P, (b) contraction of P to K_5, (c) A subdivision of $K_{3,3}$ in P

8.7. Hamiltonian Plane Graphs

An elegant necessary condition for a plane graph to be Hamiltonian was given by Grinberg [54].

Theorem 8.7.1 *Let G be a loopless plane graph having a Hamilton cycle C. Then $\sum_{i=2}^{n}(i-2)(\phi_i' - \phi_i'') = 0$, where ϕ_i' and ϕ_i'' are the numbers of faces of G of degree i contained in int C and ext C respectively.*

Proof Let E' and E'' denote the sets of edges of G contained in int C and ext C, respectively, and let $|E'| = m'$ and $|E''| = m''$. Then int C contains exactly $m' + 1$ faces (see Figure 8.27) and so

$$\sum_{i=2}^{n} \phi_i' = m' + 1 \tag{1}$$

(Since G is loopless, $\phi_1' = \phi_1'' = 0$.)

Moreover, each edge in int C is on the boundary of exactly two faces in int C, and each edge of C is on the boundary of exactly one face in int C. Hence,

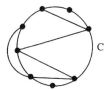

FIGURE 8.27. Graph for proof of Theorem 8.7.1

counting the edges of all the faces in int C, we get

$$\sum_{i=2}^{n} i\phi_i' = 2m' + n. \tag{2}$$

Eliminating m' from (1) and (2), we get

$$\sum_{i=2}^{n}(i - 2)\,\phi_i' = n - 2. \tag{3}$$

Similarly,

$$\sum_{i=2}^{n}(i - 2)\,\phi_i'' = n - 2. \tag{4}$$

Equations (3) and (4) give the required result. □

Grinberg's condition is quite powerful in that by using this result many plane graphs can easily be shown to be non-Hamiltonian by establishing that they do not satisfy the condition.

Example 8.7.2 The Herschel graph G of Figure 5.4 is non-Hamiltonian.

Solution G has 9 faces, and all the faces are of degree 4. Hence, if G were Hamiltonian, we must have $2(\phi_4' - \phi_4'') = 0$. This means that $\phi_4' = \phi_4''$. This is impossible, since $\phi_4' + \phi_4'' =$ number of faces of degree 4 in $G = 9$ is odd. Hence, G must be non-Hamiltonian. (In fact, it is the smallest planar non-Hamiltonian 3-connected graph.)

Exercise 7.1 Does there exist a plane Hamiltonian graph with faces of degrees 5, 7, and 8, and with just one face of degree 7?

Exercise 7.2 Prove that the Grinberg graph given below is non-Hamiltonian.

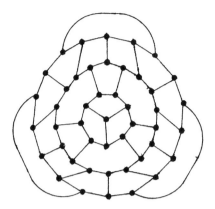

The Grinberg graph

8.8. Tait Coloring

In an attempt to solve the four-color problem, Tait considered edge colorings of 2-edge-connected cubic planar graphs. He conjectured that every such graph was 3-edge colorable. Indeed, he could prove that his conjecture was equivalent to the 4-color problem (see Theorem 8.8.1). This Tait did in 1880. He even went to the extent of giving a "proof" of the 4-color theorem using this result. Unfortunately, Tait's proof was based on the wrong assumption that any 2-edge-connected cubic planar graph is Hamiltonian. A counterexample to his assumption was given by Tutte in 1946 (65 years later). The graph given by Tutte is the graph of Figure 8.28.

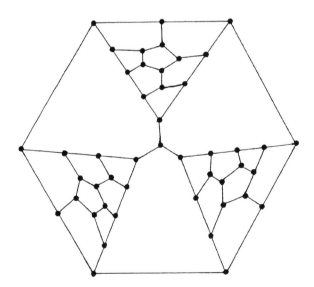

FIGURE 8.28. Tutte graph

It is a non-Hamiltonian cubic 3-connected (and therefore 3-edge-connected; see Theorem 3.3.5) planar graph. Tutte used ad hoc techniques to prove this result. (The Grinberg condition does not establish this result.)

We indicate below the proof of the fact that the Tutte graph of Figure 8.28 is non-Hamiltonian. The graphs G_1 to G_5 mentioned below are shown in Figure 8.29.

It is easy to check that there is no Hamilton cycle in the graph G_1 containing both of the edges e_1 and e_2. Now, if there is a Hamilton cycle in G_2 containing both of the edges e_1' and e_2', then there will be a Hamilton cycle in G_1 containing e_1 and e_2. Hence, there is no Hamilton cycle in G_2 containing e_1' and e_2'. In G_3-e', u and w are vertices of degree 2. Hence, if G_3-e' were Hamiltonian, then in any Hamilton cycle of G_3-e', both the edges incident to u as well as both the edges incident to w must be consecutive. This would imply that G_2 has a Hamilton cycle containing e_1' and e_2', which is not the case. Consequently, any Hamilton cycle of G_3 must contain the edge e'. It follows that there exists no Hamilton path from x to y in G_3-w. A redrawing of G_3-w is the graph G_4. It is called the "Tutte triangle." The Tutte graph (Figure 8.28) contains three copies of G_4 together with a vertex v_0. It has been redrawn as graph G_5 of Figure 8.29. Suppose G_5 is Hamiltonian with a Hamilton cycle C. If we describe C starting from v_0, it is clear that C must visit each copy of G_4 exactly once. Hence, if C enters a copy of G_4, it must exit that copy through x or y after visiting all the other vertices of that copy. But this means that there exists a Hamilton path from y to x (or from x to y) in G_4, a contradiction. Thus the Tutte graph G_5 is non-Hamiltonian.

We now give the proof of Tait's result. Recall that by Vizing-Gupta's theorem (Theorem 7.4.5), every simple cubic graph has chromatic index 3 or 4. A 3-edge coloring of a cubic planar graph is often called a *Tait coloring*.

Theorem 8.8.1   *The following statements are equivalent:*
  (i)   *All plane graphs are 4-vertex colorable.*
  (ii)   *All plane graphs are 4-face colorable.*
  (iii)   *All simple 2-edge-connected cubic planar graphs are 3-edge colorable (i.e., Tait colorable).*

Proof   *(i)* $\Rightarrow$ *(ii)*. Let G be a plane graph. Let G^* be the dual of G (see Section 8.4). Then, since G^* is a plane graph, it is 4-vertex colorable. If v^* is a vertex of G^*, and f_v is the face of G corresponding to v^*, assign to f_v the color of v^* in a 4-vertex coloring of G^*. Then, by the definition of G^*, it is clear that adjacent faces of G will receive distinct colors. (See Figure 8.30, in which f_v and f_w receive the colors of v^* and w^*, respectively). Thus, G is 4-face colorable.

(ii) $\Rightarrow$ *(iii)*. Let G be a plane embedding of a 2-edge-connected cubic planar graph. By assumption, G is 4-face colorable. Denote the four colors by $(0, 0)$, $(1, 0)$, $(0, 1)$, and $(1, 1)$, the elements of the ring $\mathbb{Z}_2 \times \mathbb{Z}_2$. If e is an edge of G that separates the faces, say f_1 and f_2, color e with the color given by the sum (in $\mathbb{Z}_2 \times \mathbb{Z}_2$) of the colors of f_1 and f_2. Since G has no cut edge, each edge is the common boundary of exactly two faces of G. This gives a 3-edge coloring

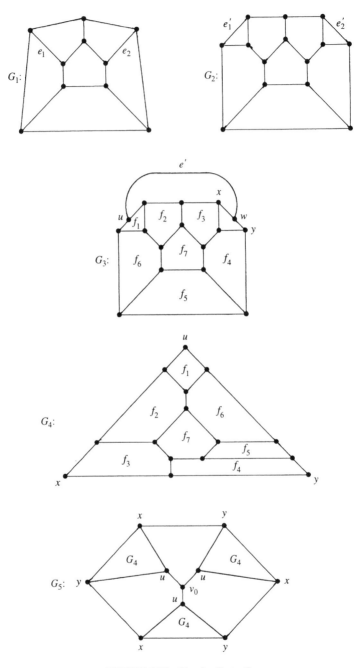

FIGURE 8.29. Graphs G_1 to G_5

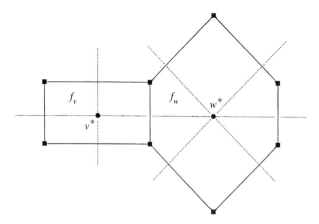

FIGURE 8.30. Graph for proof of (i) ⇒ (ii) in Theorem 8.8.1

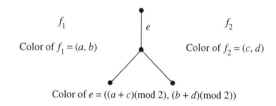

FIGURE 8.31. Graph for proof of (ii) ⇒ (iii) for Theorem 8.8.1

of G using the colors $(1, 0)$, $(0, 1)$, and $(1, 1)$, since the sum of any two distinct elements of $\mathbb{Z}_2 \times \mathbb{Z}_2$ is not $(0, 0)$. (See Figure 8.31.)

(iii) ⇒ *(i)* Let G be a planar graph. We want to show that G is 4-vertex colorable. We may assume without loss of generality that G is simple. Let $\tilde{G}$ be a plane embedding of G. Then $\tilde{G}$ is a spanning subgraph of a plane triangulation T, and hence it suffices to prove that T is 4-vertex colorable.

Let T^* be the dual of T. Then T^* is a 2-edge-connected cubic plane graph. By our assumption, T^* is 3-edge colorable using the colors, say, c_1, c_2, and c_3. Since T^* is cubic, each of the above three colors is represented at each vertex of T^*. Let T_{ij}^* be the edge-subgraph of T^* induced by the edges of T^*, which have been colored using the colors c_i and c_j. Then T_{ij}^* is a disjoint union of even cycles, and thus it is 2-face colorable. But each face of T^* is the intersection of a face of T_{12}^* and a face of T_{23}^* (see Figure 8.32). Now the 2-face colorings of T_{12}^* and T_{23}^* induce a 4-face-coloring of T^* if we assign to each face of T^* the (unordered) pair of colors assigned to the faces whose intersection is f. Since $T^* = T_{12}^* \cup T_{23}^*$, this defines a proper 4-face coloring of T^*. Thus, $\chi(G) = \chi(\tilde{G}) \le \chi(T) = \chi^*(T^*) \le 4$, and G is 4-vertex colorable. (Recall that $\chi^*(T^*)$ is the face chromatic number of T^*.) □

Exercise 8.1 Exhibit a 3-edge coloring for the Tutte graph (See Figure 8.28).

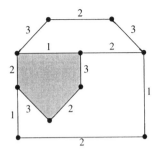

FIGURE 8.32. Graph for proof of (iii) ⇒ (i) in Theorem 8.8.1

Notes

The proof of Heawood's theorem uses arguments based on paths in which the vertices are colored alternately by two colors. Such paths are called "Kempe chains" after A. Kempe [77], who first used such chains in his "proof" of the 4CC. Even though Kempe's proof went wrong, his idea of using Kempe chains and switching the colors in such chains had been effectively exploited by Heawood [67] in proving his five-color theorem (Theorem 8.5.2) for planar graphs, as well as by Appel, Haken, and Koch [4] in settling the 4CC. As the reader might notice, the same technique had been employed in the proof of Brooks's theorem (Theorem 7.2.7). Chronologically, Francis Guthrie conceived the four-color theorem in 1852 (if not earlier). Kempe's purported "proof" of the 4CC was given in 1879, and the mistake in his proof was pointed out by Heawood in 1890. The Appel-Haken-Koch proof of the 4CC was first announced in 1976. Between 1879 and 1976, graph theory had witnessed an unprecedented growth along with the methods to tackle the 4CC. The reader who is interested in getting a detailed account of the four-color problem may consult Ore [99] and Kainen and Saaty [76].

Even though the Tutte graph of Figure 8.28 shows that not every cubic 3-connected planar graph is Hamiltonian, Tutte himself has shown that every 4-connected planar graph is Hamiltonian. (See W. T. Tutte, A theorem on planar graphs, *Trans. Amer. Math. Soc.* 82 (1956), 570–590.)

IX
Triangulated Graphs

9.0. Introduction

Triangulated graphs form an important class of graphs. They are a subclass of the class of perfect graphs and contain the class of interval graphs. They possess a wide range of applications. We describe later in this chapter an application of interval graphs in phasing the traffic lights at a road junction.

We begin with the definition of perfect graphs.

9.1. Perfect Graphs

For a simple graph G, we have the following notations:

$\chi(G)$: the chromatic number of G;

$\omega(G)$: the clique number of G (= the order of a maximum clique of G);

$\alpha(G)$: the independence number of G;

$\theta(G)$: the clique covering number of G (= the minimum number of cliques of G that cover the vertex set of G).

For instance, for the graph G of Figure 9.1, $\chi(G) = \omega(G) = 4$, and $\alpha(G) = \theta(G) = 4$.

The cliques that cover $V(G)$ are $\{1\}$, $\{2\}$, $\{3, 4, 5, 6\}$, and $\{7, 8, 9\}$. In any proper vertex coloring of G, the vertices of any clique must receive distinct colors. Hence, it is clear that $\chi(G) \geq \omega(G)$. Further, if A is any independent set of G, any clique of a clique cover of G can contain at most one vertex of A. Hence, to cover the $\alpha(G)$ vertices of a maximum independent set of G, at least $\alpha(G)$ distinct cliques of G are needed. Thus, $\theta(G) \geq \alpha(G)$.

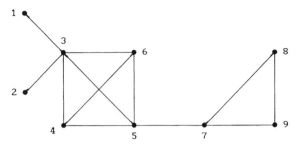

FIGURE 9.1. Graph G

If G is an odd cycle C_{2n+1}, $n \geq 2$, then $\chi(G) = 3$, $\omega(G) = 2$, $\theta(G) = n + 1$, and $\alpha(G) = n$. Hence, for such a G, $\chi(G) > \omega(G)$, and $\theta(G) > \alpha(G)$. Moreover, $A \subset V(G)$ is an independent set of vertices of G if, and only if, A induces a clique in G^c. Therefore, for any simple graph G,

$$\chi(G) = \theta(G^c), \quad \text{and}$$
$$\alpha(G) = \omega(G^c). \qquad (*)$$

Definition 9.1.1 Let G be a simple graph. Then

(i) G is χ-*perfect* if, and only if, $\chi(G[A]) = \omega(G[A])$ for every $A \subseteq V(G)$, and

(ii) G is α-*perfect* if, and only if, $\alpha(G[A]) = \theta(G[A])$ for every $A \subseteq V(G)$.

Remarks 9.1.2

1. By (*) above, it is clear that a graph is χ-perfect if, and only if, its complement is α-perfect.

2. Berge [13] conjectured that the concepts of χ-perfectness and α-perfectness are equivalent for any simple graph. This was shown to be true by Lovász [84] (and independently by Fulkerson [47]). This result is often referred to in the literature as the *perfect graph theorem*.

Theorem 9.1.3 (Perfect graph theorem)   *For a simple graph G the following statements are equivalent:*

(i)   *G is χ-perfect.*
(ii)   *G is α-perfect.*
(iii) *$\alpha(G[A])\,\omega(G[A]) \geq |A|$ for every $A \subseteq V(G)$.*

In view of the perfect graph theorem, there is no need to distinguish between α-perfectness and χ-perfectness; hence, graphs that satisfy any one of these three equivalent conditions can be referred to as merely *perfect graphs*. In particular, this means that a simple graph G is perfect if, and only if, its complement is perfect. For a proof of the perfect graph theorem, see references [52] or [84].

FIGURE 9.2. Bull graph

Remark 9.1.4   If G is perfect, by the criteria mentioned above, G cannot contain an odd hole, that is, an odd cycle C_{2n+1}, $n \geq 2$, as an induced subgraph; likewise, by (*), G cannot contain an odd antihole, that is, C^c_{2n+1}, $n \geq 2$, as an induced subgraph. Equivalently, if G is perfect, then G can contain neither C_{2n+1}, $n \geq 2$ nor its complement as an induced subgraph. The converse of this result is the celebrated "*perfect graph conjecture*" of Berge.

Remark 9.1.5   *Perfect graph conjecture:* A simple graph G is perfect if it contains neither an odd cycle of length ≥ 5 nor its complement as an induced subgraph. Although the perfect graph conjecture is still open, several classes of graphs have been shown to satisfy the conjecture. It is true for (i) $K_{1,3}$-free graphs [101]; (ii) (K_4-e)-free graphs [102]; (iii) K_4-free graphs [115]; (iv) Bull-free graphs [27] (see Figure 9.2 for the bull graph); (v) Triangulated graphs (see Theorem 9.2.7); and (vi) weakly triangulated graphs [66]. (A graph is *weakly triangulated* if it contains neither a chordless cycle of length at least five nor the complement of such a cycle as an induced subgraph. Note that any triangulated graph is weakly triangulated.)

9.2.   Triangulated Graphs

Definition 9.2.1   A simple graph G is called *triangulated* if every cycle of length at least four in G has a chord, that is, an edge joining two nonadjacent vertices of the cycle (Figure 9.3). For this reason, triangulated graphs are also called *chordal graphs* and sometimes *rigid circuit graphs*.

Remark 9.2.2   It is clear that the property of a graph being triangulated is hereditary; that is, if G is triangulated, then every induced subgraph of G is also triangulated. While discussing triangulated graphs, we need only consider simple graphs.

Definition 9.2.3   A vertex v of a graph G is a *simplicial vertex* of G if the set $N_G(v)$ of neighbors of v in G forms a clique.

Example 9.2.4   In Figure 9.3(a), the vertices u_1, u_2, u_3, and u_4 are simplicial, whereas v_1, v_2, v_3, and v_4 are not.

Triangulated graphs can be recognized by the presence of a perfect vertex elimination scheme.

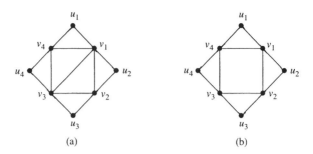

FIGURE 9.3. (a) Triangulated and (b) nontriangulated graphs

Definition 9.2.5   A *perfect vertex elimination scheme* (or, briefly, a *perfect scheme*) of a graph G is an ordering $\{v_1, v_2, \ldots, v_n\}$ of the vertex set of G in such a way that, for $1 \le i \le n - 1$, v_i is a simplicial vertex of the subgraph induced by $\{v_i, v_{i+1}, \ldots, v_n\}$ of G.

Example 9.2.6   For the graph of Figure 9.3(a), $\{u_1, u_2, u_3, u_4, v_4, v_2, v_1, v_3\}$ is a perfect scheme.

Remark 9.2.7   Any vertex of degree 1 is trivially simplicial. Hence, any tree has a perfect vertex elimination scheme. Also, any tree is trivially triangulated. It turns out that these facts can be generalized to assert that any triangulated graph has a perfect vertex elimination scheme. Based on this, Fulkerson and Gross [46] gave a "good algorithm" to test for triangu'ated graphs, namely, repeatedly locate a simplicial vertex and remove it from the graph until there is left out a single vertex and the graph is triangulated, or else at some stage no simplicial vertex exists and the graph is not triangulated.

Before we establish the above result, we need another characterization of triangulated graphs. This result is due to Hajnal and Surányi [58] and also due to Dirac [37].

Lemma 9.2.8   *A graph G is triangulated if, and only if, every minimal vertex cut of G is a clique.*

Proof   Assume that G is triangulated and that S is a minimal vertex cut of G. Let a and b be vertices in distinct components, say G_A and G_B, respectively, of $G \backslash S$. Now, every vertex x of S must be adjacent to some vertex of G_A and some vertex of G_B, since if x is adjacent to no vertex of G_A, then $G \backslash (S \backslash x)$ is disconnected and this would contradict the minimality of S. Hence, for any pair $x, y \in S$, there exist paths $P_1 : xa_1 \cdots a_r y$ and $P_2 : xb_1 \cdots b_s y$, with each $a_i \in G_A$ and each $b_j \in G_B$. Let us assume further that the a_i's and b_j's have been so chosen that these x–y paths are of least length. Then $xa_1 \cdots a_r y b_s b_{s-1} \cdots b_1 x$ is a cycle whose length is at least 4, and so it must have a chord. But such a chord cannot be of the form $a_i a_j$ or $b_k b_\ell$ in view of the minimality of the lengths of P_1 and P_2. Hence, it can be only xy. Thus, every pair x, y in S is adjacent, and S is a clique.

Conversely, assume that every minimal vertex cut of G is a clique. Let $a x b y_1 y_2 \cdots y_r a$ be a cycle C of length ≥ 4 in G. If ab were not a chord of C, denote by S a minimal vertex cut that puts a and b in distinct components of $G \backslash S$. Then S must contain x and y_j for some j. By hypothesis, S is a clique, and hence $xy_j \in E(G)$, and xy_j is a chord of C. Thus, G is triangulated.   □

Lemma 9.2.9 *Every triangulated graph G has a simplicial vertex. Moreover, if G is not complete, it has two nonadjacent simplicial vertices.*

Proof The lemma is trivial if G is either complete or if G has just two or three vertices. Assume, therefore, that G is not complete, so that G has two nonadjacent vertices a and b. Let the result be true for all graphs with fewer vertices than G. Let S be a minimal vertex cut separating a and b, and let G_A and G_B be components of $G \backslash S$ containing a and b, respectively, and with vertex sets A and B, respectively. By induction hypothesis, it follows that if $G[A \cup S]$ is not complete, it has two nonadjacent simplicial vertices. In this case, since $G[S]$ is complete (refer Lemma 9.2.8), at least one of the two simplicial vertices must be in A. Such a vertex is then a simplicial vertex of G because none of its neighbors is in B. Further, if $G[A \cup S]$ is complete, then any vertex of A is a simplicial vertex of G. In any case, we have a simplicial vertex of G in A. Similarly, we have a simplicial vertex in B. These two vertices are then nonadjacent simplicial vertices of G.   □

We are now ready to prove the second characterization theorem of triangulated graphs.

Theorem 9.2.10 *A graph G is triangulated if, and only if, it has a perfect vertex elimination scheme.*

Proof Let G be triangulated. Assume that every graph with fewer vertices than G has a perfect vertex elimination scheme. By Lemma 9.2.9, G has a simplicial vertex v. Then $G \backslash v$ has a perfect vertex elimination scheme. Then v followed by a perfect scheme of $G \backslash v$ gives a perfect scheme of G.

Conversely, assume that G has a perfect scheme, say $\{v_1, v_2, \ldots, v_n\}$. Let C be a cycle of length ≥ 4 in G. Let j be the first suffix with $v_j \in V(C)$. Then $V(C) \subseteq G[\{v_j, v_{j+1}, \ldots, v_n\}]$ and, since v_j is simplicial in $G[\{v_{j+1}, \ldots, v_n\}]$, the neighbors of v_j in C are adjacent, and hence C has a chord. Thus G is triangulated.   □

Theorem 9.2.11 *A triangulated graph is perfect.*

Proof Let G be a triangulated graph. Assume that the theorem is true for all graphs having fewer vertices than G. If G is disconnected, we can consider each component of G individually. So assume that G is connected. By Lemma 9.2.9,

G contains a simplicial vertex v. Let u be a vertex adjacent to v in G. Since v is simplicial in G (and so in $G-u$), $\theta(G-u) = \theta(G)$. By the induction hypothesis, $\theta(G-u) = \alpha(G-u)$, and hence $\theta(G) = \theta(G-u) = \alpha(G-u) \leq \alpha(G)$. This, together with the fact that $\theta(G) \geq \alpha(G)$, implies that $\theta(G) = \alpha(G)$, and the proof is complete since by the induction assumption, for any proper subset A of $V(G)$, the subgraph $G[A]$ is triangulated and therefore perfect. □

9.3. Interval Graphs

One of the special classes of triangulated graphs is the class of interval graphs.

Definition 9.3.1   An *interval graph* G is the intersection graph of a family of intervals of the real line. This means that for each vertex v of G, there corresponds an interval $J(v)$ of the real line such that $uv \in E(G)$ if, and only if, $J(u) \cap J(v) \neq \emptyset$.

Figure 9.4 displays a graph G and its interval representation.

Remarks 9.3.2

1. Interval graphs occur in a natural manner in various applications. In genetics, the Benzer model [11] deals with the conditions under which two subsets of the fine structure inside a gene overlap. In fact, one can tell when they overlap on the basis of mutation data. Is this overlap information consistent with the hypothesis that the fine structure inside the gene is linear? The answer is "yes" if the graph defined by the overlap information is an interval graph.

2. It is clear that the intervals may be taken as either open or closed. Clearly, the cycle C_4 is not an interval graph. In fact, if $V(C_4) = \{a, b, c, d\}$ and if ab, bc, cd, and da are the edges of C_4, then $J(a) \cap J(b) \neq \emptyset, J(b) \cap J(c) \neq \emptyset, J(c) \cap J(d) \neq \emptyset$, and $J(d) \cap J(a) \neq \emptyset$ imply that either $J(a) \cap J(c) \neq \emptyset$ or $J(b) \cap J(d) \neq \emptyset$ (i.e., $ac \in E(G)$ or $bd \in E(G)$), which is not the case. Hence, an interval graph cannot contain C_4 as an induced subgraph. For a similar reason, it can be checked that the graph H of Figure 9.5 is not an interval graph.

3. Recall that an *orientation* of a graph G is an assignment of a direction to each edge of G. Hence, an orientation of G converts G into a directed graph. As mentioned in Chapter II, an orientation is *transitive* if when (a, b) and (b, c) are arcs in the orientation, then (a, c) is also an arc in the orientation.

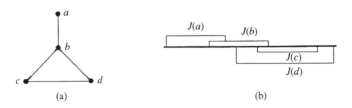

FIGURE 9.4. (a) Graph G; (b) its interval representation

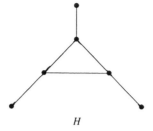

H

FIGURE 9.5. Example of a graph that is not an interval graph

Lemma 9.3.3   *If G is an interval graph, then G^c has a transitive orientation.*

Proof   Let $J(a)$ denote the interval that represents the vertex a of the interval graph G. Let $ab \in E(G^c)$, and $bc \in E(G^c)$ so that $ab \notin E(G)$ and $bc \in E(G)$. Hence, $J(a) \cap J(b) = \emptyset$, and $J(b) \cap J(c) = \emptyset$. Now, introduce an orientation for the edges of G^c by orienting an edge ab of G^c from a to b whenever $J(a)$ lies to the left of $J(b)$. Then, if $J(b)$ lies to the left of $J(c)$, $J(a)$ also lies to the left of $J(c)$, and hence, whenever (a, b) and (b, c) are arcs in the defined orientation, arc (a, c) also belongs to the orientation. Thus, G^c has a transitive orientation. $\square$

Gilmore and Hoffman [50] have shown that the above two properties characterize interval graphs.

Theorem 9.3.4   *A graph G is an interval graph if, and only if, G does not contain C_4 as an induced subgraph and G^c admits a transitive orientation.*

Proof   We have just seen the necessity of these two conditions. We now prove their sufficiency. Assume that G has no induced C_4 and that G^c has a transitive orientation. We look at the set of maximal cliques of G and introduce a linear ordering on it. If A and B are two distinct maximal cliques of G, there exist vertices $a \in A$ and $b \in B$ such that $ab \notin E(G)$ and therefore $ab \in E(G^c)$. (Otherwise, each vertex of A is adjacent to every vertex of B, and therefore $G[A \cup B]$ would be a clique properly containing A and B, a contradiction, since A and B are maximal.) If ab has the orientation from a to b in the transitive orientation of G^c, we set $A < B$. This ordering is well defined in that if $a' \in A$ and $b' \in B$, with $a'b' \in E(G^c)$, then $a'b'$ must be oriented from a' to b' in G^c (see Figure 9.6).

To see this, first assume that $a \neq a'$ and $b \neq b'$ and that $a'b'$ is oriented from b' to a' in G^c. Then at least one of the edges ab' and $a'b$ must be an edge of G^c. Otherwise, the edges $aa', a'b, bb'$, and $b'a$ induce a C_4 in G, a contradiction. Suppose, then, $a'b \in E(G^c)$. Then if $a'b$ is oriented from a' to b in G^c, by the transitivity of the orientation in G^c, $b'b \in E(G^c)$, a contradiction. A similar argument applies when ba', ab', or $b'a$ is an oriented arc of G^c. The cases when $a = a'$ or $b = b'$ can also be treated similarly. Thus, if one arc of G^c goes from A to B, then all the arcs between A and B go from A to B in G^c. Since the number

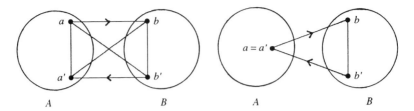

FIGURE 9.6. Graph for proof of first condition of Theorem 9.3.4

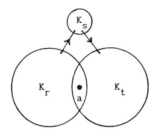

FIGURE 9.7. Ordering of maximal cliques

of maximal cliques of G is finite, this gives a linear ordering of the set of maximal cliques of G, say $K_1, K_2, \ldots, K_p$.

We now claim that if a vertex a of G belongs to K_r and K_t, then it also belongs to K_s, where $K_r < K_s < K_t$ (see Figure 9.7).

First note that there exists some vertex b in K_s such that b is nonadjacent to a. If not, $K_s \vee \{a\}$ would be a clique properly containing K_s, a contradiction. But then, since $K_r < K_s$, the edge ab of G^c must be oriented from a to b. But $a \in K_t$, and this means that $K_t < K_s$, a contradiction. Thus, $a \in K_s$ as well.

In $\{1, 2, \ldots, p\}$, let i be the smallest and j be the greatest numbers such that $a \in K_i$ and $a \in K_j$. We now define the interval $J(a) = [i, j]$. It is clear that $J(a) \cap J(b) \neq \emptyset$ if, and only if, there exists a positive integer k such that $k \in J(a) \cap J(b)$. But this can happen if, and only if, both a and b are in K_k (i.e., $ab \in E(G)$). Thus, G is an interval graph. $\square$

9.4. Bipartite Graph $B(G)$ of a Graph G

Given a graph G, we define the associated bipartite graph $B(G)$ as follows: Let $V(G) = \{v_1, v_2, \ldots, v_n\}$. Corresponding to $V(G)$, take disjoint sets $X = \{x_1, x_2, \ldots, x_n\}$ and $Y = \{y_1, y_2, \ldots, y_n\}$ and form the bipartite graph $B(G)$ by taking X and Y as sets of the bipartition of the vertex set of $B(G)$. Adjacency in $B(G)$ is defined by setting $x_i y_i \in E(B(G))$ for every i, $1 \leq i \leq n$, and for $i \neq j$, x_i is adjacent to y_j in $B(G)$ if, and only if, $v_i v_j \in E(G)$ (Figure 9.8).

Our next theorem relates the chordal nature of a graph G with that of the bipartite graph $B(G)$. Since a bipartite graph has no odd cycles and a 4-cycle of a bipartite

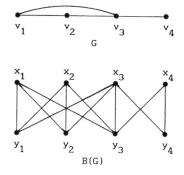

FIGURE 9.8. Bipartite graph $B(G)$ of G

graph cannot have a chord, a *bipartite graph* is defined to be *chordal* if each of its cycles of length at least 6 has a chord.

Theorem 9.4.1   *If the bipartite graph $B(G)$ formed out of G is chordal then G is chordal.*

Proof   Let $C = v_1 v_2 \cdots v_m v_1$ be any cycle of length ≥ 4 in G. Then $x_1 y_1 x_2 y_2 \ldots$ $x_m y_m x_1$ is a cycle C_1 of length ≥ 8 in $B(G)$. Since $B(G)$ is chordal, C_1 has a chord. But this chord can only be of the form $x_i y_j$, $i \neq j$ (for $x_i y_i \in E(C_1)$ for each i). Since $x_i y_j$ is an edge of $B(G)$, $v_i v_j \in E(G)$, and so $v_i v_j$ is a chord of C. Hence, G is chordal.   □

9.5.   Circular Arc Graphs

Circular arc graphs are similar to interval graphs except that the $J(a)$'s are now taken to be arcs of a particular circle. Consider an interval graph G. Since the number of intervals $J(a)$, $a \in V(G)$, is finite, there are real numbers m and M such that $J(a) \subseteq (m, M)$ for every $a \in V(G)$. Consequently, identification of m and M (i.e., conversion of the closed interval $[m, M]$ into a circle by the identification of m and M) makes G a circular arc graph. Thus, every interval graph is a circular arc graph. Clearly, the converse is not true. However, if there exists a point p on the circle that does not belong to any arc $J(a)$, then the circle can be cut at p and the circular arc graph can be made into an interval graph.

9.6.   Exercises

6.1   If e is an edge of a cycle of a triangulated graph G, show that e belongs to a triangle of G.

6.2   What are the simplicial vertices of the triangulated graph of Figure 9.3(a)?

6.3   Give a perfect elimination scheme for the triangulated graph of Figure 9.3(a).

6.4   If v is a simplicial vertex of a triangulated graph G, and $vu \in E(G)$, prove that
$\theta(G{-}u) = \theta(G)$.

6.5   Let $t(G)$ denote the smallest positive integer k such that G^k is triangulated. Determine
$t(C_n), n \geq 4$.

6.6   Prove: G and G^c are triangulated if, and only if, G does not contain C_4, C_4^c, or C_5 as
an induced subgraph. Hence, or otherwise, show that $C_n^c, n \geq 5$ is not triangulated.

6.7   Prove that $L(G)$ is triangulated if, and only if, every block of G is either K_2 or K_3.
Hence, show that the line graph of a tree is triangulated.

6.8   Let $K(G)$ and $L(G)$ denote, respectively, the clique graph and the line graph of a
graph G. ($K(G)$ is defined as the intersection graph of the family of maximal cliques
of G; i.e., the vertices of $K(G)$ are the maximal cliques of G and two vertices of
$K(G)$ are adjacent in $K(G)$ if, and only if, the corresponding maximal cliques of G
have a nonempty intersection.) Then prove or disprove:
(i) G is triangulated $\Rightarrow$ $K(G)$ is triangulated;
(ii) $K(G)$ is triangulated $\Rightarrow$ G is triangulated;
(iii) $L(G)$ is triangulated $\Rightarrow$ G is triangulated;
(iv) G is triangulated $\Rightarrow$ $L(G)$ is triangulated.

6.9   Show by means of an example that an even power of a triangulated graph need not
be triangulated.

6.10   Prove the following by means of a counterexample: G is chordal need not imply
that $B(G)$ is chordal.

6.11   Draw the interval graph of the family of intervals below and display a transitive
orientation for G^c.

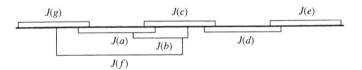

6.12   If G is cubic and if G does not contain an odd cycle of length at least five as an
induced subgraph, then prove that G is perfect (Hint: Use Brooks's theorem.)

6.13   Show that every bipartite graph is perfect.

6.14   For a bipartite graph G, prove that $\chi(G^c) = \omega(G^c)$.

6.15   Give an example of a triangulated graph that is not an interval graph.

6.16   Give an example of a perfect graph that is not triangulated.

6.17   Show that a 2-connected triangulated graph with at least four vertices is locally connected. Hence, show that a 2-connected triangulated $K_{1,3}$-free graph is Hamiltonian.
(See reference [97].)

6.18   Show by means of an example that a 2-connected triangulated graph need not be Hamiltonian.

6.19   Show that the line graph of a 2-edge connected triangulated graph is Hamiltonian.

6.20   Give an example of a circular arc graph that is not an interval graph.

6.21* Show that a graph G is perfect if, and only if, every induced subgraph G' of G contains an independent set that meets all the maximum cliques of G'.

9.7.   Phasing of Traffic Lights at a Road Junction

We present here an application of interval graphs to the problem of phasing of traffic lights at a road junction. The problem is to install traffic lights at a road junction in such a way that traffic flows smoothly and efficiently at the junction.

We take a specific example and explain how our problem could be tackled. Figure 9.9 displays the various traffic streams, namely $a, b, \ldots, g$ that meet at the Main Guard Gate road junction at Tiruchirapalli, Tamil Nadu (India).

Certain traffic streams may be termed compatible if their simultaneous flow would not result in any accidents. For instance, in Figure 9.9, streams a and d are compatible, whereas b and g are not. The phasing of lights should be such that when the green lights are on for two streams, they should be compatible. We suppose that the total time for the completion of green and red lights during one cycle is two minutes.

We form a graph G whose vertex set consists of the traffic streams in question and we make two vertices of G adjacent if, and only if, the corresponding streams are compatible. This graph is the compatibility graph corresponding to the problem in question. The compatibility graph of Figure 9.9 is shown in Figure 9.10.

We take a circle and assume that its perimeter corresponds to the total cycle period, namely, 120 seconds. We may think that the duration when a given traffic stream gets a green light corresponds to an arc of this circle. Hence, two such arcs of the circle can overlap only if the corresponding streams are compatible. The

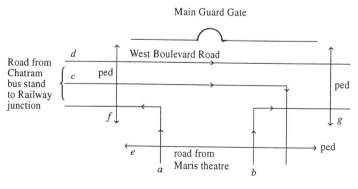

FIGURE 9.9. Traffic streams (ped = pedestrian crossing)

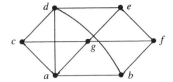

FIGURE 9.10. Compatibility graph of Figure 9

FIGURE 9.11. A green light assignment

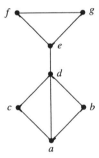

FIGURE 9.12. Intersection graph for Figure 9.11

resulting circular arc graph may not be the compatibility graph because we do not demand that two arcs intersect whenever they correspond to compatible flows. (There may be two compatible streams but they need not get a green light at the same time.) However, the intersection graph H of this circular arc graph will be a spanning subgraph of the compatibility graph.

The efficiency of our phasing may be measured by minimizing the total red light time during a traffic cycle, that is, the total waiting time for all the traffic streams during a cycle. For the sake of concreteness, we may assume that at the time of starting, all lights are red. This would ensure that H is an interval graph (see the last sentence of Section 9.5 on circular arc graphs).

Figure 9.11 gives a feasible green light assignment whose corresponding intersection graph H is given in   Figure 9.12. The maximal cliques of H are $K_1 = \{a, b, d\}$, $K_2 = \{a, c, d\}$, $K_3 = \{d, e\}$, and $K_4 = \{e, f, g\}$. Since H is an interval graph, by Theorem 9.3.4, H^c has a transitive orientation. A transitive orientation of H^c is given in Figure 9.13.

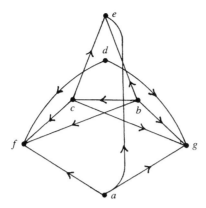

FIGURE 9.13. Transitive orientation of H^c

Since (b, c), (c, e), and (d, f) are arcs of H^c, and since $b \in K_1$, $c \in K_2$, $d \in K_3$ and $e \in K_4$, etc., we have

$$K_1 < K_2 < K_3 < K_4$$

in the consecutive ordering of the maximal cliques of H. Each clique K_i, $1 \le i \le 4$, corresponds to a phase during which all streams in that clique receive green lights. We then start a given traffic stream with a green light during the first phase in which it appears, and keep it green until the last phase in which it appears. Because of the consecutiveness of the ordering of the phases K_i, this gives an arc on the clock circle. In phase 1, traffic streams a, b, and d receive a green light; in phase 2, a, c, and d receive a green light, and so on.

Suppose we assign to each phase K_i a duration d_i. Our aim is to determine the d_i's (≥ 0) so that the total waiting time is minimum. Further, we may assume that the minimum green light time for any stream is 20 seconds. Traffic stream a gets a red light when the phases K_3 and K_4 receive a green light. Hence a's total red light time is $d_3 + d_4$. Similarly, the total red light times of traffic streams b, c, d, e, f, and g, respectively, are $d_2 + d_3 + d_4$; $d_1 + d_3 + d_4$; d_4; $d_1 + d_2$; $d_1 + d_2 + d_3$; and $d_1 + d_2 + d_3$. Therefore, the total red light time of all the streams in one cycle is $Z = 4d_1 + 4d_2 + 4d_3 + 3d_4$. Our aim is to minimize Z subject to $d_i \ge 0$, $1 \le i \le 4$, and $d_1 + d_2 \ge 20$, $d_1 \ge 20$, $d_2 \ge 20$, $d_1 + d_2 + d_3 \ge 20$, $d_3 + d_4 \ge 20$, $d_4 \ge 20$, and $d_1 + d_2 + d_3 + d_4 = 120$. (The condition $d_1 + d_2 \ge 20$ signifies that the green light time that stream a receives, namely, the sum of the green light times of phases K_1 and K_2, is at least 20. A similar reasoning applies to the other inequalities. The last condition gives the total cycle time.) An optimal solution to this problem is $d_1 = 80$, $d_2 = 20$, $d_3 = 0$, and $d_4 = 20$ and min $Z = 480$ (in seconds). But this is not the end of our problem. There are other possible circular arc graphs. Figures 9.14(a), (b) give another feasible green light arrangement and its corresponding intersection graph. With respect to this graph, min $Z = 500$ seconds. Thus, we have to exhaust all possible circular arc graphs and then take the least of all the minima thus obtained. The phasing that corresponds to this least

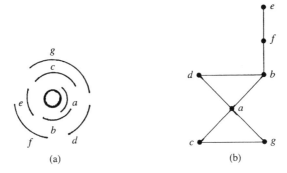

FIGURE 9.14. (a) Another green light arrangement; (b) corresponding intersection graph

value would then be the best phasing of the traffic lights. (For the above particular problem, it is clear that the minimum value is 480 seconds.)

Notes

Exercise 6.9 shows that an even power of a triangulated graph need not be triangulated. However, an odd power of a triangulated graph is triangulated [6]. Moreover, if G^k is triangulated, then so is G^{k+2} [82], and consequently, if G and G^2 are triangulated, then so are all the powers of G.

  Four books that give a very good account of perfect graphs are references [13], [14], [15], and [52]. In addition to the classes of perfect graphs mentioned in Section 9.1, there are also other known classes of perfect graphs, for instance, wing-triangulated graphs and, more generally, strict quasi-parity graphs. For details, see reference [69]. Our discussion on phasing of traffic lights is based on Roberts [108], which also contains some other applications of perfect graphs.

X
Applications

10.0. Introduction

In this chapter, we consider some immediate applications of certain graph theoretic results to day-to-day life problems.

10.1. The Connector Problem

Problem 1. Various cities in a country are to be linked by means of roads. Given the various possibilities of connecting the cities and the costs involved, what is the most economical way of laying roads so that in the resulting road network any two cities are connected by a chain of roads? Similar problems involve designing railroad networks, or water-line transports.

Problem 2. A layout for a housing colony in a city is to be prepared. Various locations of the colony are to be linked by roads. Given the various possibilities of linking the locations and their costs, what is the minimum-cost layout so that any two locations are connected by a chain of roads?

Problem 3. A layout for the electrical wiring of a building is to be prepared. Given the costs of the various possibilities, what is the minimum-cost layout?

These three problems are particular cases of a graph-theoretical problem known as the "connector problem."

Definition 10.1.1 Let G be a graph. To each edge e of G, we associate a nonnegative number $w(e)$ called its *weight*. The resulting graph is a *weighted graph*. If H is a subgraph of G, the sum of the weights of the edges of H is called the *weight of H*. In particular, the sum of the weights of the edges of a path is called the *weight of the path*.

We shall now concentrate on Problem 1. Problems 2 and 3 can be similarly considered. Let G be a graph constructed with the set of cities as its vertex set. An edge of G corresponds to a road link between cities. The cost of construction of a road link is the weight of its corresponding edge. Then a minimum weight spanning tree of G provides the most economical layout for the road network.

We present two algorithms, Kruskal's algorithm and Prim's algorithm, for determining a minimum-weight spanning tree in a connected weighted graph. We can assume, without loss of generality, that the graph is simple, for, since no loop can be an edge of a spanning tree, we can discard all loops. Also, since we are interested in determining a minimum-weight spanning tree, we can retain, from a set of multiple edges having the same ends, an edge with the minimum weight, and we can discard all the others.

First we describe Kruskal's algorithm [79].

10.2.   Kruskal's Algorithm

Let G be a simple connected weighted graph. The three steps of the algorithm are as follows:

Step 1: Choose an edge e_1 with its weight $w(e_1)$ as small as possible.

Step 2: If the edges $e_1, e_2, \ldots, e_i, i \geq 1$, have already been chosen, choose e_{i+1} from the set $E \setminus \{e_1, e_2, \ldots, e_i\}$ such that

(i)   the subgraph induced by the edge set $\{e_1, e_2, \ldots, e_{i+1}\}$ is acyclic, and

(ii) $w(e_{i+1})$ is as small as possible subject to (i).

Step 3: Stop when step 2 cannot be implemented further.

When the graph is not weighted, we can give the weight 1 to each of its edges and then apply the algorithm. The algorithm then gives an acyclic subgraph with as many edges as possible, that is, a spanning tree of G.

Illustration   The distances in miles between some of the Indian cities connected by Indian Airlines are given in Table 10.1.

TABLE 10.1. *Mileage between Indian cities*

	Mumbai	Hyderabad	Nagpur	Calcutta	New Delhi	Chennai
Mumbai (M)						
Hyderabad (H)	385					
Nagpur (N)	425	255				
Calcutta (Ca)	1035	740	679			
New Delhi (D)	708	773	531	816		
Chennai (Ch)	644	329		860	1095	

Determine a minimum-cost operational system so that every city is connected to every other city. Assume that the cost of operation is directly proportional to the distance.

Let G be a graph with the set of cities as its vertex set. An edge corresponds to a pair of cities for which ticketed mileage is indicated. The ticketed mileage is the weight of the corresponding edge (see Figure 10.2).

The required operation system demands a minimum cost spanning tree of G. We shall apply Kruskal's algorithm and determine such a system. The following is a sequence of edges selected according to the algorithm.

$$HN, HCh, HM, ND, NCa.$$

The corresponding spanning tree is shown in Figure 10.3, and its weight is $255 + 329 + 385 + 531 + 679 = 2179$.

We next describe Prim's algorithm [105].

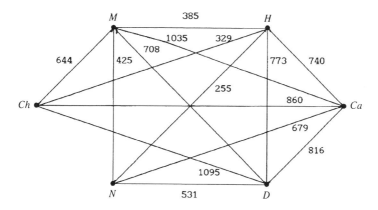

FIGURE 10.2. Graph of mileage between cities

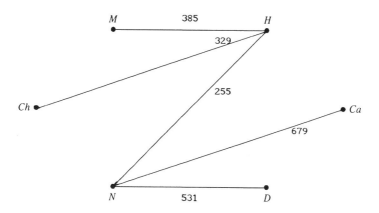

FIGURE 10.3. A minimum-weight spanning tree of the graph of Figure 10.2

10.3.   Prim's Algorithm

Let G be a simple connected weighted graph having n vertices. Let the vertices of G be labeled as: $v_1, v_2, \ldots, v_n$. Let $W = W(G) = (w_{ij})$ be the weight matrix of G. That is, W is the $n \times n$ matrix with

  (i) $w_{ii} = \infty$, for $1 \leq i \leq n$,
  (ii) $w_{ij} = w_{ji} =$ the weight of the edge (v_i, v_j), if v_i and v_j are adjacent,
  (iii) $w_{ij} = w_{ji} = \infty$, if v_i and v_j are nonadjacent.

The algorithm constructs a minimum-cost spanning tree.

Step 1: Start with v_1. Connect v_1 to v_k, where v_k is a nearest vertex to v_1 (v_k is nearest to v_1 if $v_1 v_k$ is an edge with minimum possible weight). The vertex v_k could be easily determined by observing the matrix W. Actually, v_k is a vertex corresponding to which the entry in row 1 of W is minimum.

Step 2: Having chosen v_k, let $v_i \neq v_1$ or v_k be a vertex corresponding to the smallest entry in rows 1 and k put together. Then v_i is the vertex "nearest" to the edge subgraph defined by the edge $v_1 v_k$. Connect v_i to v_1 or v_k, according to whether the entry is in the first row or kth row. Suppose it is, say, in the kth row; then it is the (k, i)th entry of W.

Step 3: Consider the edge subgraph defined by the edge set $\{v_1 v_k, v_k v_i\}$. Determine the nearest neighbor to the set of vertices $\{v_1, v_k, v_i\}$.

Step 4: Continue the process until all the n vertices have been connected by $(n - 1)$ edges. This results in a minimum-cost spanning tree.

Illustration   Consider the weighted graph G shown in Figure 10.2. The weight matrix W of G is

$$
W : \begin{array}{c} \\ M \\ H \\ N \\ Ca \\ D \\ Ch \end{array}
\begin{array}{cccccc}
M & H & N & Ca & D & Ch \\
\infty & 385 & 425 & 1035 & 708 & 644 \\
385 & \infty & 255 & 740 & 773 & 329 \\
425 & 255 & \infty & 679 & 531 & \infty \\
1035 & 740 & 679 & \infty & 816 & 860 \\
708 & 773 & 531 & 816 & \infty & 1095 \\
644 & 329 & \infty & 860 & 1095 & \infty
\end{array}
$$

In row M (i.e., in the row corresponding to the city M, namely, Mumbai), the smallest weight is 385, which occurs in column H. Hence, join M and H. Now, after omitting columns M and H, 255 is the minimum weight in the rows M and H put together. It occurs in row H and column N. Hence, join H and N. Now, omitting columns M, H, and N, the smallest number in the rows M, H, and N put together is 329, and it occurs in row H and column Ch, so join H and Ch. Again, the smallest entry in rows M, H, N, and Ch not belonging to the corresponding columns is 531, and it occurs in row N and column D. So join N and D. Now, the least entry in rows M, H, N, D, and Ch not belonging to the corresponding columns is 679, and it occurs in row N and column Ca. So join N and Ca. This construction gives the same minimum-weight spanning tree of Figure 10.3.

Remark   In each iteration of Prim's algorithm, a subtree of a minimum-weight spanning tree is obtained, whereas the subgraph constructed in any step of Kruskal's algorithm is just a subgraph of a minimum-weight spanning tree.

10.4.   Shortest-Path Problems

A manufacturing concern has a warehouse at location X and the market for the product at another location Y. Given the various routes of transporting the product from X to Y and the cost of operating them, what is the most economical way of transporting the materials? This problem can be tackled by graph theory. All such optimization problems come under a type of graph theoretic problem known as "shortest path problems." Three types of shortest-path problems are well-known:
Let G be a connected weighted graph.

1.   Determine a shortest path, that is, a minimum-weight path between two specified vertices of G.
2.   Determine a set of shortest paths between all pairs of vertices of G.
3.   Determine a set of shortest paths from a specified vertex to all other vertices of G.

We consider only the first problem. The other two problems are similar. We describe Dijkstra's algorithm [32] for determining the shortest path between two specified vertices. Once again, it is clear that in shortest-path problems, we could restrict ourselves to simple connected weighted graphs.

Dijkstra's algorithm: Let G be a simple connected weighted graph having vertices $v_1, v_2, \ldots, v_n$. Let s and t be two specified vertices of G. We want to determine a shortest path from s to t. Let W be the weight matrix of G. Dijkstra's algorithm allots weights to the vertices of G. At each stage of the algorithm, some vertices have permanent weights and others have temporary weights.

To start with, the vertex s is allotted the permanent weight 0 and all other vertices the temporary weight ∞. In each iteration of the algorithm, one new vertex is allotted a permanent weight by the following rules:

Rule 1: If v_j is a vertex that has not yet been allotted a permanent weight, determine, for each vertex v_i that had already been allotted a permanent weight,

$$\alpha_{ij} = \min\{\text{old weight of } v_j, (\text{old weight of } v_i) + w_{ij}\}.$$

Let $w_j = \text{Min}_i\, \alpha_{ij}$. Then w_j is a new temporary weight of v_j not exceeding the previous temporary weight.

Rule 2: Determine the smallest among the w_j's. If this smallest weight is at v_k, w_k becomes the permanent weight of v_k. In case there is a tie, any one vertex is taken for allotting a permanent weight.

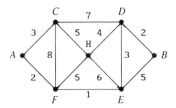

FIGURE 10.4. Graph G for shortest-path problem

The algorithm stops when the vertex t gets a permanent weight.

It is clear from the algorithm that the permanent weight of each vertex is the shortest weighted distance from s to that vertex. The shortest path from s to t is constructed by working backward from the terminal vertex t. Let P be a shortest path and p_i a vertex of P. The weight of p_{i-1}, the vertex immediately preceding p_i on P, is such that the weight of the edge $p_{i-1} p_i$ equals the difference in permanent weights of p_i and p_{i-1}.

We present below an illustrative example. Let us find the shortest path from A to B in the graph of Figure 10.4.

We present the various iterations of the algorithm by arrays of weights of the vertices, one consisting of weights before iteration and another after it. Temporary weights will be enclosed in squares and the permanent weights enclosed in double squares. The steps of the algorithm for determining the shortest path from vertex A to vertex B in graph G of Figure 10.4 are given in Table 10.5.

In our example, B is the last vertex to get a permanent weight. Hence, the algorithm stops after Iteration 6, in which B is allotted the permanent weight. However, the algorithm may be stopped as soon as vertex B gets the permanent weight.

The shortest distance from A to B is 8. A shortest path is $AFEDB$ with weight 8.

TABLE 10.5. *Steps of algorithm for shortest path from A to B*

	A	B	C	D	E	F	H
Iteration 0	⟦0⟧	∞	∞	∞	∞	∞	∞
Iteration 1	0	∞	3	∞	∞	2	∞
Iteration 2	0	∞	3	∞	3	2	7
Iteration 3	0	∞	3	10	3	2	7
Iteration 4	0	8	3	6	3	2	7
Iteration 5	0	8	3	6	3	2	7
Iteration 6	0	8	3	6	3	2	7

10.5.   Timetable Problem

The timetable problem was briefly considered in Section 7.4. Following are some examples of such a problem.

Problem 1.   In a social health check-up scheme, specialist physicians are to visit various health centers. Given the places each physician has to visit and also the time interval of his visit, how can we fit in an itinerary? The assumption is that each health center can accommodate only one doctor at a time and each doctor is available at only one place at a given time.

Problem 2.   Mobile laboratories are to visit various schools in a city. Given the places each lab has to visit and also the time interval (period) of visits in a day, how can we fit in a timetable for the laboratories?

Problem 3.   In an educational institution, as is well known, teachers have to meet various classes. Given the various classes each teacher has to meet in a day, how can we fit in a timetable? It is presumed that a teacher can teach only one class at a time and that each class could be taught by only one teacher at a time.

We shall now discuss problem 3. Let $x_1, x_2, \ldots, x_n$ denote the teachers, and $y_1, y_2, \ldots, y_m$ the classes. Let t_{ij} denote the number of periods for which teacher x_i has to meet class y_j. How can we draw up a timetable? If there are constraints on the availability of class rooms, what is the minimum number of periods required to implement a timetable? If the number of periods in a day is specified, what is the minimum number of rooms required to implement the timetable? All these problems could be analyzed by using a suitable graph.

Let $G(X, Y)$ be a bipartite graph formed with $X = \{x_1, x_2, \ldots, x_n\}$ and $Y = \{y_1, y_2, \ldots, y_m\}$ as the bipartition and in which there are t_{ij} parallel edges with x_i and y_j as their common ends. Then a teaching assignment for a period determines a matching in the bipartite graph G. Conversely, any matching in G corresponds to a teaching assignment for a period. The edges of G could be partitioned into Δ edge-disjoint matchings (see Theorem 7.4.3). Corresponding to the Δ matchings, a Δ-period timetable can be drawn up.

Let T be the total number of periods to be met by all teachers put together. Then, on an average, T/Δ classes are to be met per period. Hence, at least $\lceil T/\Delta \rceil$ rooms are necessary to implement a Δ-period timetable. We present below a method for drawing up such a timetable. For this, we need Lemma 10.5.1.

Lemma 10.5.1   *Let M and N be disjoint matchings of a graph G with $|M| > |N|$. Then there are disjoint matchings M' and N' of G with $|M'| = |M| - 1$ and $|N'| = |N| + 1$ and with $M' \cup N' = M \cup N$.*

Proof   Consider the subgraph $H = G[M \cup N]$. Each component of H is either an even cycle or a path with edges alternating between M and N. Since $|M| > |N|$, some path component P of H must have its initial and terminal edges in M. Let $P = v_0 e_1 v_1 e_2 v_2 \cdots e_{2r+1} v_{2r+1}$.

Now set

$$M' = (M \setminus \{e_1, e_3, \ldots, e_{2r+1}\}) \cup \{e_2, e_4, \ldots, e_{2r}\}$$

and

$$N' = (N \setminus \{e_2, e_4, \ldots, e_{2r}\}) \cup \{e_1, e_3, \ldots, e_{2r+1}\}.$$

Then M' and N' are disjoint matchings of G satisfying the conditions of the lemma.

$\square$

Theorem 10.5.2 *If G is a bipartite graph (with m edges), and if $t \geq \Delta$, then there exist t disjoint matchings $M_1, M_2, \ldots, M_t$ of G such that*

$$E = M_1 \cup M_2 \cup \cdots \cup M_t$$

and, for $1 \leq i \leq t$,

$$\lfloor m/t \rfloor \leq |M_i| \leq \lceil m/t \rceil.$$

Proof By Theorem 7.4.3, $\chi' = \Delta$. Hence, $E(G)$ can be partitioned into Δ matchings $M'_1, M'_2, \ldots, M'_\Delta$. Hence, for $t \geq \Delta$, there exist disjoint matchings $M'_1, M'_2, \ldots, M'_t$, where $M'_i = \emptyset$ for $\Delta + 1 \leq i \leq t$ such that

$$E = M'_1 \cup M'_2 \cup \cdots \cup M'_t.$$

Now, repeatedly apply Lemma 10.5.1 to pairs of matchings that differ by more than one in size. This would eventually result in matchings $M_1, M_2, \ldots, M_t$ of G satisfying the condition stated in the theorem. $\square$

Coming back to our timetable problem, if the number of rooms available, say r, is less than T/Δ (so that $T/r > \Delta$), then the number of periods is to be correspondingly increased. Hence, starting with an edge partition of $E(G)$ into matchings $M'_1, M'_2, \ldots, M'_\Delta$, we apply Lemma 10.5.1 repeatedly to get an edge partition of $E(G)$ into disjoint matchings $M_1, M_2, \ldots, M_{\lceil T/r \rceil}$. This partition gives a $\lceil T/r \rceil$-period timetable that uses r rooms.

Illustration The teaching assignments of five teachers, x_1, x_2, x_3, x_4, x_5, in the Mathematics Department of a particular college are given by the following array:

	I Year y_1	II Year y_2	III Year y_3	IV Year y_4
x_1	1	2	—	—
x_2	1	1	1	—
x_3	1	—	—	2
x_4	—	—	1	—
x_5	—	—	1	1

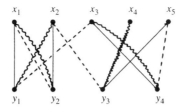

FIGURE 10.6. Bipartite graph corresponding to problem 1

TABLE 10.7. *Timetable*

		Period		
		I	II	III
Teacher:	x_1	y_1	y_2	y_2
	x_2	y_2	y_3	y_1
	x_3	y_4	y_1	y_4
	x_4	—	—	y_3
	x_5	y_3	y_4	—

The bipartite graph G corresponding to the above problem is shown in Figure 10.6.

Each of the sets of edges drawn by the solid lines, wavy lines, and dashed lines gives a matching in G. The three matchings cover the edges of G. Hence, they can be the basis of a 3-period timetable. The corresponding timetable is given in Table 10.7.

In each period, four classes are to be met. Hence, at least four rooms are needed to implement this timetable. Here, $\Delta = 3$ and $T = 12$. Consequently, G could be covered by three matchings each containing $\lfloor 12/3 \rfloor$ or $\lceil 12/3 \rceil$ edges, or exactly 4 edges. This gives the edge-partition $M' = \{M'_1, M'_2, M'_3\}$, where

$$M'_1 = \{x_1 y_1, x_2 y_2, x_3 y_4, x_5 y_3\},$$

$$M'_2 = \{x_1 y_2, x_2 y_3, x_3 y_1, x_5 y_4\},$$

$$M'_3 = \{x_1 y_2, x_2 y_1, x_3 y_4, x_4 y_3\}.$$

Now, take $M'' = \{M'_1, M'_2, M'_3, M'_4 = \varnothing\}$, and apply Lemma 10.5.1. This gives an edge-partition $M = \{M_1, M_2, M_3, M_4\}$, where $M_1 = \{x_1 y_1, x_2 y_2, x_3 y_4\}$; $M_2 = \{x_1 y_2, x_2 y_3, x_5 y_4\}$; $M_3 = \{x_2 y_1, x_3 y_4, x_4 y_3\}$; and $M_4 = \{x_5 y_3, x_3 y_1, x_1 y_2\}$. The above partition yields a four-period timetable using three rooms.

10.6. Application to Social Psychology

The study of applications of graph theory to social psychology was initiated by Cartwright, Harary, and Norman [62] when they were at the Research Center for Group Dynamics at the University of Michigan during the 1960s.

Group dynamics is the study of social relationships between people within a particular group. The graphs that are commonly used to study these relationships are signed graphs. A *signed graph* is a graph G with sign $+$ or $-$ attached to each of its edges. An edge of G is *positive* (respectively *negative*) if the sign attached to it is $+$ (respectively $-$). A positive sign between two persons u and v would mean that u and v are "related," that is, they share the same social trait under consideration. A negative sign would indicate the opposite. The social trait may be "same political ideology," "friendship," "likes certain social customs," and so on.

A group of people with such relations between them is called a *social system*. A social system is called *balanced* if any two of its people have a positive relation between them, or it is possible to divide the group into two subgroups so that any two persons in the same subgroup have positive relation between them while two persons of the different subgroups have negative relation between them. This, of course, means that if both u and v have negative relation to w, then u and v must have positive relation between them.

In consonance with a balanced social system, a balanced signed graph G is defined as a graph in which the vertex set V can be partitioned into two subsets $V_i, i = 1, 2$, one of which may be empty, so that any edge in each $G[V_i]$ is positive while any edge between V_1 and V_2 is negative.

For example, the signed graph G_1 of Figure 10.8(a) is balanced (take $V_1 = \{u_1, u_2\}$ and $V_2 = \{u_3, u_4, u_5\}$). However, the signed graph G_2 of Figure 10.8(b) is not balanced. (If G_2 were balanced, and $v_4 \in$ (say) V_1, then $v_5 \in V_2$ and $v_0 \in V_1$, and therefore v_0v_5 should be a negative edge, which is not the case.)

A *path* or *cycle* in a signed graph is *positive* if it has an even number of $-$ signs; otherwise, the path is negative. The following characterization of balanced signed graphs is due to Harary [62].

Theorem 10.6.1 (Harary [62])   *A signed graph S is balanced if, and only if, the paths between any two vertices of S either are all positive paths or are all negative paths.*

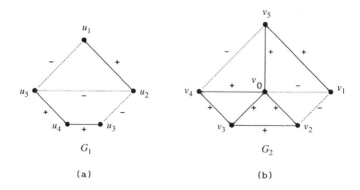

FIGURE 10.8. (a) Balanced and (b) unbalanced graph

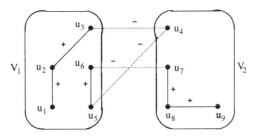

A negative u_1-u_9 path

FIGURE 10.9. A negative u_1–u_9 path

(a)   (b)

FIGURE 10.10. Path P with (a) $uw \in P$ and (b) $uw \notin P$

Proof   Let S be a balanced signed graph with (V_1, V_2) being the parition of the vertex set $V(S)$ of S. Let $P: u = u_1 u_2 \cdots u_n = v$ be a path in S. Without loss of generality, we may assume that $u_1 \in V_1$. Then, as we traverse along P from u_1, we will continue to remain in V_1 until we traverse along a negative edge. Recall that a negative edge joins a vertex of V_1 to a vertex of V_2. Hence, if P contains an odd number of negative edges, that is, if P is negative, then $v \in V_2$, whereas if P is positive, $v \in V_1$ (see Figure 10.9). It is clear that every u–v path in S must have the same sign as P.

Conversely, assume that S is a signed graph with the property that between any two vertices of S the paths are either all positive or all negative. We prove that S is balanced. We may assume that S is a connected graph. Otherwise, we can prove the result for the components, and if (V_i, V_i'), $1 \le i \le \omega$, are the partitions of the vertex subsets of the (signed) components, then $[\cup_{i=1}^{\omega} V_i, \cup_{i=1}^{\omega} V_i']$ is a partition of $V(S)$ of the requisite type.

So we assume that S is connected. Let v be any vertex of S. Denote by V_1 the set of all vertices u of S that are connected to v by positive paths of S, and let $V_2 = V(S) \backslash V_1$. Then no edge both of whose end vertices are in V_1 can be negative. Suppose, for instance, $u \in V_1$, $w \in V_1$, and edge uw is negative. Let P be any v–w path in S. Since $w \in V_1$, P is a positive path. If $uw \in P$, (Figure 10.10(a)), then P–(uw) is a negative v–u path in S, contradicting the choice of $u \in V_1$. If $uw \notin P$, (Figure 10.10(b)), then $P \cup (wu)$ is a negative v–u path in S, again a contradiction. For a similar reason, no edge both of whose end vertices are in V_2 can be negative, and no edge with one end in V_1 and the other end in V_2 can be positive. Thus (V_1, V_2) is a partition of $V(S)$ of the required type, which ensures that S is balanced. $\square$

10.7. Exercises

7.1 For graph G, determine two distinct minimum-weight spanning trees using

    (i) Kruskal's algorithm,

    (ii) Prim's algorithm.

What is the weight of such a tree? Also determine a minimum-weight s–t path using Dijkstra's algorithm.

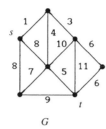

G

7.2 Apply Prim's algorithm to the illustrative example given in Section 10.3 by starting from the third row of the weight matrix.

7.3 If G is a connected weighted graph in which no two edges have the same weight, show that G has a unique minimum spanning tree.

7.4 Draw a period timetable for the following teaching assignment:

	C_1	C_2	C_3	C_4	C_5	C_6	C_7
T_1	1	—	1	—	1	—	—
T_2	—	1	—	1	1	—	—
T_3	—	1	1	—	—	1	—
T_4	1	—	1	—	—	—	1
T_5	1	1	—	—	—	1	1

Draw up a timetable using four rooms. Also draw up another timetable if only three rooms are available.

7.5 Prove that a signed graph S is balanced if, and only if, every cycle of S is positive.

7.6 Let S be a social system in which every two individuals dislike each other. Is S balanced?

7.7 Let S be a signed tree. Is S balanced?

Notes

In this chapter, we have touched only the fringes of graph algorithms by mentioning three important algorithms. Graph theory has witnessed a spectacular growth, and its importance in recent years has grown mainly due to its practical applications (as already mentioned in Chapter I) in that several practical problems can be formulated as graph problems and solved using graph algorithms. A number of books deal with graph algorithms. Of these, special mention can be made of references [1], [49], [52], [53], [93], and [100]. The books [48] and [100] give a nice account of time complexity of graph algorithms.

List of Symbols

$A(D)$	the set of arcs of D; 49
$B(G)$	the bipartite graph of graph G; 299
c	the capacity function of a network; 87
$c(a)$	the capacity of arc a; 87
C_n	the cycle of length n; 19
$\mathrm{cap}\,K$	the sum of the capacities of the arcs in K; 89
$cl(G)$	the closure of G; 172
$d_G(v)$	the degree of the vertex v in G; 14
$d(v)$	the degree of the vertex v in a graph; 14
$(d_1, d_2, \ldots, d_n)$	the degree sequence of a graph; 15
$d(u, v)$	the length of a shortest u–v path (respectively directed path) in a graph; 20 (respectively digraph; 64)
$d_D^+(v)$	the outdegree of v in D; 50
$d^+(v)$	the outdegree of v in a digraph; 50
$d_D^-(v)$	the indegree of v in D; 51
$d^-(v)$	the indegree of v in a digraph; 51
$d_D(v)$	the degree of v in D; 51
D	a directed graph or digraph; 49
D_n	the dihedral group of order $2n$; 46
$\mathrm{diam}(G)$	the diameter of G; 47
$E(G)$	the edge set of G; 3
E	the edge set of a graph; 5
$e(v)$	the eccentricity of vertex v; 110
$f^+(S)$	$f([S, \bar{S}])$, where $S \subseteq V(D)$; 88

$f^-(S)$	$f([\bar{S}, S])$, where $S \subseteq V(D)$; 88
$f(G)$	the number of faces of a planar graph G; 244
$f(G; \lambda)$	the chromatic polynomial of G; 229
f_{uv}	$f((u, v))$; 88
G	a graph; 3
G^c	the complement of a simple graph G; 9
$G(D)$	the underlying graph of D; 49
$G(X, Y)$	a bipartite graph G with bipartition (X, Y); 9
$G[S]$	the subgraph of G induced by the subset S of $V(G)$; 11
$G[E']$	the subgraph of G induced by the subset E' of $E(G)$; 11
$G + uv$	the supergraph of G obtained by adding the new edge uv; 11
$G - v$	the subgraph of G obtained by deleting the vertex v; 13
$G - S$	the subgraph of G obtained by the deletion of the vertices in S; 13
$G - e$	the subgraph of G obtained by deleting the edge e; 13
$G - E'$	the subgraph of G obtained by the deletion of the edges in E'; 13
$G_1 \cup G_2$	the union of the two graphs G_1 and G_2; 37
$G_1 + G_2$	the sum of the two graphs G_1 and G_2; 37
$G_1 \cap G_2$	the intersection of the two graphs G_1 and G_2; 37
$G_1 \vee G_2$	the join of the two graphs G_1 and G_2; 37
$G_1 \times G_2$	the Cartesian product of the graph G_1 with the graph G_2; 38
$G_1[G_2]$	the composition or lexicographic product of the graph G_1 with the graph G_2; 39
$G_1 \circ G_2$	the normal product or the strong product of the graph G_1 with the graph G_2; 40
$G_1 \otimes G_2$	the tensor product or the Kronecker product of the graph G_1 with the graph G_2; 41
$G \circ e$	the graph obtained from G by contracting the edge e; 116
G^*	the canonical dual of the plane graph G; 256
G_4	Grötszch graph; 213
G^k	the k-th power of G; 42
I_D	the incidence map of D; 49
I_G	the incidence map of G; 3
K_n	the complete graph on n vertices; 7
$K_{p,q}$	the complete bipartite graph with part sizes p and q; 9

$K_{1,q}$	the star of size q; 9
$K(G)$	the clique graph of G; 299
$L(G)$	the line graph of the graph G; 29
$m(G)$	the size of G = the number of edges in G; 5
m	the size (= the number of edges) of a graph; 5
$N_G(v)$	the open neighborhood of the vertex v in G; 4
$N(v)$	the open neighborhood of the vertex v in a graph; 4
$N_G[v]$	the closed neighborhood of the vertex v in G; 4
$N[v]$	the closed neighborhood of the vertex v in a graph; 4
$n(G)$	the order of G = the number of vertices of G; 5
n	the order of a graph; 5
$N_D^+(v)$	the set of outneighbors of v in D; 50
$N^+(v)$	the set of outneighbors of v in a digraph; 50
$N_D^-(v)$	the set of inneighbors of v in D; 50
$N^-(v)$	the set of inneighbors of v in a digraph; 50
N	a network; 87
$N(S)$	the neighbor set of S in a graph; 139
$o(G)$	the number of odd components of G; 144
P	the Petersen graph; 8
P_n	the path on n vertices; 19
P^{-1}	the inverse of the path P; 19
Q_n	the n-cube; 136
rG	the sum of r copies of the graph G; 37
$r(G)$	the radius of graph G; 111
s	the symmetric group of degree n; 27
S_n	the source of a network; 87
$[S, S']$	the set of all arcs having their tails in S and heads in S' in the case of directed graphs; 55 (the set of all edges having one end in S and the other end in S' in the case of undirected graphs; 67)
$s(v)$	the score of the vertex v in a tournament; 61
$(s(v_1), s(v_2), \ldots, s(v_n))$	the score vector of a tournament with vertex set $\{v_1, v_2, \ldots, v_n\}$; 61
t	the sink of a network; 87
$v_0 e_1 v_1 e_2 v_2 \ldots e_r v_r$	a (v_0, v_r) walks in a graph; 18
V	the vertex set of a graph; 5
$V(D)$	the set of vertices of D; 49
$V(G)$	the vertex set of G; 3
val f	the value of the flow $f = f^+(s) - f^-(s) = f^-(t) - f^+(t)$; 89
W_n	$C_n \vee K_1$, the wheel with n spokes; 37

$\cong$	is isomorphic to; 7
$\alpha(G)$	the independence number of G; 129
$\alpha'(G)$	the cardinality of a maximum matching of G; 130
$\beta(G)$	the covering number of G; 129
$\beta'(G)$	the cardinality of a minimum edge covering of G; 130
$\Gamma(G)$	the group of automorphisms of the graph G; 25
$\delta(G)$	the minimum degree of G; 14
δ	the minimum degree of a graph; 14
$\Delta(G)$	the maximum degree of G; 14
Δ	the maximum degree of a graph; 14
$\phi_1 \circ \phi_2$	the composition of the mappings ϕ_1 and ϕ_2 (ϕ_2 followed by ϕ_1); 26
$\lambda(G)$	the edge connectivity of G; 73
λ	the edge connectivity of a graph; 73
$\lambda_c G$	the cyclical edge connectivity of G; 85
$\kappa(G)$	the vertex connectivity of G; 73
κ	the vertex connectivity of a graph; 73
$\theta(G)$	the clique covering number of G = the minimum number of cliques of G that cover the vertex set of G; 285
$\tau(G)$	the number of spanning trees of G; 116
$\omega(G)$	the clique number of G = the order of a maximum clique of G; 285
$\omega(G)$	the number of components of G; 20
$\chi(G)$	the chromatic number of G; 199
$\chi'(G)$	the edge chromatic number or chromatic index of G; 215

References

[1] A. V. Aho, J. E. Hopcroft, and J. D. Ullman, *The Design and Analysis of Computer Algorithms*, Addison-Wesley, Reading, Mass., 174.

[2] K. Appel and W. Haken, Every planar map is four colorable: Part I—Discharging, *Illinois J. Math.* **21** (1977), 429–490.

[3] K. Appel and W. Haken, The solution of the four-color-map problem, *Scientific American* **237(4)** (1977), 108–121.

[4] K. Appel, W. Haken, and J. Koch, Every planar map is four colorable: Part II—Reducibility, *Illinois J. Math.* **21** (1977), 491–567.

[5] R. Aravamudhan and B. Rajendran (personal communication).

[6] R. Balakrishnan and P. Paulraja, Powers of chordal graphs, *J. Austral. Math. Soc. Ser. A* **35** (1983), 211–217.

[7] R. Balakrishnan and P. Paulraja, Line graphs of subdivision graphs, *J. Combin. Info. and Sys. Sci.* **10** (1985), 33–35.

[8] R. Balakrishnan and P. Paulraja, Chordal graphs and some of their derived graphs, *Congressus Numerantium*, **53** (1986), 33–35.

[9] M. Behzad, G. Chartrand, and L. Lesniak-Foster, *Graphs and Digraphs*, Prindle, Weber & Schmidt Int. Series, Boston, Mass., 1979.

[10] L. W. Beineke, On derived graphs and digraphs, *in Beiträge zur Graphentheorie*, eds., H. Sachs, H. J. Voss, and H. Walther, Teubner, Leipzig, 1968, 17–23.

[11] S. Benzer, On the topology of the genetic fine structure, *Proc. Nat. Acad. Sci., USA*, **45** (1959), 1607–1620.

[12] C. Berge, Färbung von Graphen, deren sämtliche bzw. deren ungerade Kreise starr sind, *Wiss Z Martin-Luther-Univ. Halle-Wittenberg Math. Natur. Reihe* (1961), 114–115.

[13] C. Berge, *Graphs and Hypergraphs*, North-Holland, London, 1973.

[14] C. Berge and V. Chvátal, Topics on perfect graphs, *Ann. Discrete Math.* **21** (1984).

[15] C. Berge, *Graphs*, Elsevier Science Publishers B.V., North-Holland, London, 1991.

[16] G. Birkhoff and D. Lewis, Chromatic polynomials, *Trans. Amer. Math. Soc.* **60** (1946), 355–451.

[17] J. A. Bondy, Pancyclic graphs, *J. Combin. Theory Ser. B* **11** (1971), 80–84.

[18] J. A. Bondy and V. Chvátal, A method in graph theory, *Discrete Math.* **15** (1976), 111–135.

[19] J. A. Bondy and U. S. R. Murty, *Graph Theory with Applications*, The MacMillan Press Ltd., 1976.

[20] J. A. Bondy and F. Y. Halberstam, Parity theorems for paths and cycles in graphs, *J. Graph Theory* **10** (1986), 107–115.

[21] B. Bollobás, *Extremal Graph Theory*, Academic Press, London, 1978.

[22] R. L. Brooks, On colouring the nodes of a network, *Proc. Cambridge Philos. Soc.* **37** (1941), 194–197.

[23] A. Cayley, On the theory of analytical forms called trees, *Philos. Mag.* **13** (1857); "Mathematical Papers," Cambridge **3** (1891), 242–246.

[24] G. Chartrand and C. E. Wall, On the Hamiltonian index of a graph, *Studia Sci. Math. Hungar.* **8** (1973), 43–48.

[25] G. Chartrand and O. R. Ollermann, *Applied and Algorithmic Graph Theory*, International Series in Pure and Applied Mathematics, McGraw-Hill, Inc., New York, 1993.

[26] V. Chvátal, On Hamilton's ideals, *J. Combin. Theory, Ser. B* **12** (1972), 163–168.

[27] V. Chvátal and N. Sbihi, Bull-free Berge graphs are perfect, *Graphs and Combinatorics* **3** (1987), 127–139.

[28] V. Chvátal and P. Erdös, A note on Hamiltonian circuits, *Discrete Math.* **2** (1972), 111–113.

[29] J. Clark and D. A. Holton, *A First Look at Graph Theory*, World Scientific Publishing Co., Inc., Teaneck, NJ, 1991.

[30] G. Demoucron, Y. Malgrange, and R. Pertuiset, Graphes planaires: reconnaissance et construction de représentations planaires topologiques, *Rev. Française Recherche Opérationnelle* **8** (1964), 33–47.

[31] B. Descartes, Solution to advanced problem no. 4526, *Amer. Math. Monthly* **61** (1954), 352.

[32] E. W. Dijkstra, A note on two problems in connexion with graphs, *Numerische Math.* **1** (1959) 269–271.

[33] G. A. Dirac, Généralisations du théorème de Menger, *C. R. Acad. Sci. Paris* **250** (1960), 4252–4253.

[34] G. A. Dirac, A property of 4-chromatic graphs and some remarks on critical graphs, *J. London Math. Soc.* **27** (1952), 85–92.

[35] G. A. Dirac, Some theorems on abstract graphs, *Proc. London Math. Soc.* **2** (1952), 69–81.

[36] G. A. Dirac, 4-chrome Graphen und vollständige 4-Graphen, *Math. Nachr.* **22** (1960), 51–60.

[37] G. A. Dirac, On rigid circuit graphs–cut sets–coloring, *Abh. Math. Sem. Univ. Hamburg* **25** (1961), 71–76.

[38] I. Fáry, On straight line representation of planar graphs, *Acta Sci. Math. Szeged* **11** (1948), 229–233.

[39] S. Fiorini and R. J. Wilson, Edge-Colourings of graphs, *in Research Notes in Mathematics* **16**, Pitman, London, 1971.

[40] H. Fleischner, Elementary proofs of (relatively) recent characterizations of Eulerian graphs, *Discrete Applied Math.* **24** (1989), 115–119.

[41] H. Fleischner, Eulerian graphs and related topics, *Annals of Disc. Math.* **45**, North-Holland, New York, 1990.

[42] L. R. Ford, Jr., and D. R. Fulkerson, Maximal flow through a network, *Canad. J. Math.* **8** (1956), 399–404.

[43] L. R. Ford, Jr., and D. R. Fulkerson, *Flows in Networks*, Princeton University Press, Princeton, 1962.

[44] J.-C. Fournier, Colorations des arêtes d'un graphe, *Cahiers du CERO* **15** (1973), 311–314.

[45] J.-C. Fournier, Demonstration simple du theoreme de Kuratowski et de sa forme duale, *Discr. Math.* **31** (1980), 329–332.

[46] D. R. Fulkerson and O. A. Gross, Incidence matrices and interval graphs, *Pacific J. Math.* **15** (1965), 835–855.

[47] D. R. Fulkerson, Blocking and anti-blocking pairs of polyhedra, *Math. Programming* **1** (1971), 168–194.

[48] M. R. Garey and D. S. Johnson, *Computers and Intractability: A Guide to the Theory of NP-Completeness*, W.H. Freeman & Co., San Francisco, 1979.

[49] A. Gibbons, *Algorithmic Graph Theory*, Cambridge University Press, Cambridge, 1985.

[50] P. C. Gilmore and A. J. Hoffman, A characterization of comparability graphs and interval graphs, *Canad. J. Math.* **16** (1964), 539–548.

[51] W. D. Goddard, G. Kubicki, O. R. Oellermann, and S. L. Tian, On multipartite tournaments, *J. Combin. Theory Ser. B* **52** (1991), 284–300.

[52] M. C. Golumbic, *Algorithmic Graph Theory and Perfect Graphs*, Academic Press, New York, 1980.

[53] R. J. Gould, *Graph Theory*, The Benjamin/Cummings Publishing Company, Inc., Menlo Park, Calif., 1988.

[54] È. Ja. Grinberg, Plane homogeneous graphs of degree three without Hamiltonian circuits (Russian), *Latvian Math. Yearbook* **4** (1968), 51–58.

[55] J. L. Gross and T. W. Tucker, *Topological Graph Theory*, John Wiley & Sons, New York, 1987.

[56] R. P. Gupta, The chromatic index and the degree of a graph, *Notices Amer. Math. Soc.* **13** (1966), Abstract 66T-429.

[57] A. Gyárfás, J. Lehel, J. Nešetril, V. Rödl, R. H. Schelp, and Z. Tuza, Local *k*-colorings of graphs and hypergraphs, *J. Combin. Theory Ser. B* **43** (1987), 127–139.

[58] A. Hajnal and J. Surányi, Über die Auflösung von Graphen in Vollständige Teilgraphen, *Ann. Univ. Sci. Budapest, Eötvös Sect. Math.* **1** (1958), 113–121.

[59] P. Hall, On representatives of subsets, *J. London Math. Soc.* **10** (1935), 26–30.

[60] M. Hall, Jr., *Combinatorial Theory*, Blaisdell, Waltham, Mass., 1967.

[61] F. Harary, *Graph Theory*, Addison-Wesley, Reading, Mass., 1969.

[62] F. Harary, R. Z. Norman, and D. Cartwright, *Structural Models: An Introduction to the Theory of Directed Graphs*, J. Wiley & Sons, New York, 1965.

[63] F. Harary and C. St. J. A. Nash-Williams, On Eulerian and Hamiltonian graphs and line graphs, *Canad. Math. Bull.* **8** (1965), 701–710.

[64] F. Harary and W. T. Tutte, A dual form of Kuratowski's theorem, *Canad. Math. Bull.* **8** (1965), 17–20.

[65] F. Harary and E. M. Palmer, *Graphical Enumeration*, Academic Press, New York, 1973.

[66] R. B. Hayward, Weakly triangulated graphs, *J. Combin. Theory Ser. B* **39** (1985), 200–209.

[67] P. J. Heawood, Map colour theorems, *Quart. J. Math.* **24** (1890), 332–338.

[68] D. A. Holton and J. Sheehan, *The Petersen Graph*, Australian Math. Soc. Lecture Series 7, Cambridge University Press, Cambridge, 1993.

[69] S. Hougardy, V. B. Le, and A. Wagler, Wing-triangulated graphs are perfect, *J. Graph Theory* **24** (1997), 25–31.

[70] F. Jaeger, A note on sub-Eulerian graphs, *J. Graph Theory* **3** (1979), 91–93.

[71] F. Jaeger, Nowhere-zero flow problems, in *Selected Topics in Graph Theory III*, eds. L. W. Beineke and R. J. Wilson, Academic Press, London (1988), 71–95.

[72] T. R. Jensen and B. Toft, *Graph Coloring Problems*, Wiley-Interscience Series in Discrete Mathematics and Optimization, John Wiley & Sons, New York, 1995.

[73] C. Jordan, Sur les assemblages de lignes, *J. Reine Agnew. Math.* **70** (1869), 185–190.

[74] Jüng, Zu einem Isomorphiesatz von Whitney für Graphen, *Math. Ann.* **164** (1966), 270–271.

[75] M. Jünger, W. R. Pulleyblank, and G. Reinelt, On partitioning the edges of graphs into connected subgraphs, *J. Graph Theory* **9** (1985), 539–549.

[76] P. C. Kainen and T. L. Saaty, *The Four-Color Problem (Assaults and Conquest)*, Dover Publications, Inc., New York, 1977.

[77] A. Kempe, On the geographical problem of the four colours, *Amer. J. Math.* **2** (1879), 193–200.

[78] P. A. Kilpatrick, Tutte's first colour-cycle conjecture, Ph.D. thesis, Cape Town, 1975.

[79] J. B. Kruskal, Jr., On the shortest spanning subtree of a graph and the travelling salesman problem, *Proc. Amer. Math. Soc.* **7** (1956), 48–50.

[80] S. Kundu, Bounds on the number of disjoint spanning trees, *J. Combin. Theory Ser. B* **17** (1974), 199–203.

[81] C. Kuratowski, Sur le problème des courbes gauches en topologie, *Fund. Math.* **15** (1930), 271–283.

[82] R. Laskar and D. Shier, On powers and centers of chordal graphs, *Discrete Applied Math.* **6** (1983), 139–147.

[83] Linda M. Lesniak, Neighborhood unions and graphical properties, *in Proceedings of the Sixth Quadrennial International Conference on the Theory and Applications of Graphs; Graph Theory, Combinatorics and Applications*, Western Michigan University, eds. Y. Alavi, G. Chartrand, O. R. Oellermann, and A. J. Schwenk, John Wiley & Sons, New York, 1991, 783–800.

[84] L. Lovász, Normal hypergraphs and the perfect graph conjecture, *Disc. Math.* **2** (1972), 253–267.

[85] L. Lovász, Three short proofs in graph theory, *J. Combin. Theory Ser. B* **19** (1975), 111–113.

[86] L. Lovász and M. D. Plummer, Matching theory, *Ann. Discrete Mathematics*, 29, North-Holland Mathematics Studies **121**, 1986.

[87] T. A. McKee, Recharacterizing Eulerian: Intimations of new duality, *Discrete Math.* **51** (1984), 237–242.

[88] K. Menger, Zur allgemeinen Kurventheorie, *Fund. Math.* **10** (1927), 96–115.

[89] J. W. Moon, On subtournaments of a tournament, *Canad. Math. Bull.* **9** (1966), 297–301.

[90] J. W. Moon, Various proofs of Cayley's formula for counting trees, *in A Seminar on Graph Theory*, ed. F. Harary, Holt, Rinehart and Winston, Inc., New York, 1967, 70–78.

[91] J. W. Moon, *Topics on Tournaments*, Holt, Rinehart and Winston Inc., New York, 1968.

[92] J. Mycielski, Sur le coloriage des graphs, *Colloq. Math.* **3** (1955), 161–162.

[93] Narasingh-Deo, *Graph Theory with Applications to Engineering and Computer Science*, Prentice Hall, Englewood Cliffs, N.J., 1974.

[94] C. St. J. A. Nash-Williams, Edge-disjoint spanning trees of finite graphs, *J. London Math. Soc.* **36** (1961), 445–450.

[95] L. Nebesky, On the line graph of the square and the square of the line graph of a connected graph, *Casopis. Pset. Mat.* **98** (1973), 285–287.

[96] E. A. Nordhaus and J. W. Gaddum, On complementary graphs, *Amer. Math. Monthly* **63** (1956), 175–177.

[97] D. J. Oberly and D. P. Sumner, Every connected, locally connected nontrivial graph with no induced claw is Hamiltonian, *J. Graph Theory* **3** (1979), 351–356.

[98] O. Ore, Note on Hamilton circuits, *Amer. Math. Monthly* **67** (1960), 55.

[99] O. Ore, *The Four-Colour Problem*, Academic Press, New York, 1967.

[100] C. H. Papadimitriou and K. Steiglitz, *Combinatorial Optimization: Algorithms and Complexity*, Prentice-Hall, Upper Saddle River, N.J., 1982.

[101] K. R. Parthasarathy and G. Ravindra, The strong perfect-graph conjecture is true for $K_{1,3}$-free graphs, *J. Combin. Theory Ser. B* **21** (1976), 212–223.

[102] K. R. Parthasarathy and G. Ravindra, The validity of the strong perfect-graph conjecture for (K_4–e)-free graphs, *J. Combin. Theory Ser. B* **26** (1979), 98–100.

[103] K. R. Parthasarathy, *Basic Graph Theory*, Tata McGraw-Hill Publishing Company Limited, New Delhi, 1994.

[104] J. Petersen, Die Theorie der regulären Graphen, *Acta Math.* **15** (1891), 193–220.

[105] R. C. Prim, Shortest connection networks and some generalization, *Bell System Techn. J.* **36** (1957), 1389–1401.

[106] D. K. Ray-Chaudhuri and R. M. Wilson, Solution of Kirkman's schoolgirl problem, *in Proceedings of the Symposium on Mathematics* **19**, American Mathematical Society, Providence, Rhode Island, 1971, 187–203.

[107] L. Rédei, Ein kombinatorischer Satz., *Acta. Litt. Sci. Szeged* **7** (1934), 39–43.

[108] F. S. Roberts, *Graph Theory and its Applications to Problems in Society*, CBMS-NSF Regional Conference Series in Mathematics, SIAM, Philadelphia, 1978.

[109] E. Sampathkumar, A characterization of trees, J. Karnatak University (Science), (1987).

[110] J. P. Serre, *Trees*, Springer-Verlag, New York, 1980.

[111] D. P. Sumner, Graphs with 1-factors, *Proc. Amer. Math. Soc.* **42** (1974), 8–12.

[112] S. Toida, Properties of an Euler graph, *J. Franklin Inst.* **295** (1973), 343–346.

[113] N. Trinajstic, *Chemical Graph Theory—Volume I*, CRC Press, Inc., Boca Raton, Fla., 1983.

[114] N. Trinajstic, *Chemical Graph Theory—Volume II*, CRC Press, Inc., Boca Raton, Fla., 1983.

[115] A. Tucker, The validity of Perfect Graph Conjecture for K_4-free graphs, *in Topics on Perfect Graphs,* eds. C. Berge and V. Chvátal, **21**, 1984, 149–157.

[116] W. T. Tutte, The factorization of linear graphs, *J. London Math. Soc.* **22** (1947), 107–111.

[117] W. T. Tutte, On the problem of decomposing a graph into n connected factors, *J. London Math. Soc.* **36** (1961), 221–230.

[118] V. G. Vizing, On an estimate of the chromatic class of a p-graph (in Russian). *Diskret. Analiz.* **3** (1964), 25–30.

[119] K. Wagner, Über eine Eigenschaft der ebenen Komplexe, *Math. Ann.* **114** (1937), 570–590.

[120] D. J. A. Welsh, *Matroid Theory*, Academic Press, London, 1976.

[121] H. Whitney, Congruent graphs and the connectivity of graphs, *Amer. J. Math.* **54** (1932), 150–168.

[122] H. P. Yap, *Some Topics in Graph Theory*, London Math. Soc. Lecture Note Series, **108**, Cambridge University Press, Cambridge, 1986.

[123] A. A. Zykov, On some properties of linear complexes (in Russian), *Math. Sbornik N. S.* 24 (1949), 163–188; *Amer. Math. Soc. Translation* **79** (1952).

Index

Universitext   *(continued)*

Meyer: Essential Mathematics for Applied Fields
Mines/Richman/Ruitenburg: A Course in Constructive Algebra
Moise: Introductory Problems Course in Analysis and Topology
Morris: Introduction to Game Theory
Polster: A Geometrical Picture Book
Porter/Woods: Extensions and Absolutes of Hausdorff Spaces
Ramsay/Richtmyer: Introduction to Hyperbolic Geometry
Reisel: Elementary Theory of Metric Spaces
Rickart: Natural Function Algebras
Rotman: Galois Theory
Rubel/Colliander: Entire and Meromorphic Functions
Sagan: Space-Filling Curves
Samelson: Notes on Lie Algebras
Schiff: Normal Families
Shapiro: Composition Operators and Classical Function Theory
Simonnet: Measures and Probability
Smith: Power Series From a Computational Point of View
Smoryski: Self-Reference and Modal Logic
Stillwell: Geometry of Surfaces
Stroock: An Introduction to the Theory of Large Deviations
Sunder: An Invitation to von Neumann Algebras
Tondeur: Foliations on Riemannian Manifolds
Wong: Weyl Transforms
Zhang: Matrix Theory: Basic Results and Techniques
Zong: Sphere Packings
Zong: Strange Phenomena in Convex and Discrete Geometry